363 .728 N964j

Nuclear waste management :

NUCLEAR WASTE MANAGEMENT

THE OCEAN
ALTERNATIVE

Pergamon Titles of Related Interest

Auer ENERGY AND THE DEVELOPING NATIONS
Colglazier THE POLITICS OF NUCLEAR WASTE
Constans MARINE SOURCES OF ENERGY
Gabor BEYOND THE AGE OF WASTE, 2nd ed.
Mossaver-Rahmani ENERGY POLICY IN IRAN
Neff THE SOCIAL COSTS OF SOLAR ENERGY
Stewart TRANSITIONAL ENERGY POLICY 1980-2030:
Alternative Nuclear Technologies

Related Journals*

ANNALS OF NUCLEAR RESEARCH
DEEP-SEA RESEARCH
NUCLEAR AND CHEMICAL WASTE MANAGEMENT
OCEAN ENGINEERING
WATER RESEARCH

*Free specimen copies available upon request.

PERGAMON
POLICY
STUDIES ON ENERGY

Nuclear Waste Management
The Ocean Alternative

Edited by
Thomas C. Jackson
The Oceanic Society

Sponsored by: The Oceanic Society
in cooperation with
The Georgetown University
Law Center Institute for
International and Foreign Trade Law and
The Center for Law and Social Policy

Pergamon Press
NEW YORK • OXFORD • TORONTO • SYDNEY • PARIS • FRANKFURT

Pergamon Press Offices:

U.S.A.   Pergamon Press Inc., Maxwell House, Fairview Park,
                     Elmsford, New York 10523, U.S.A.

U.K.   Pergamon Press Ltd., Headington Hill Hall,
                     Oxford OX3 OBW, England

CANADA   Pergamon Press Canada Ltd., Suite 104, 150 Consumers Road,
                     Willowdale, Ontario M2J 1P9, Canada

AUSTRALIA   Pergamon Press (Aust.) Pty. Ltd., P.O. Box 544,
                     Potts Point, NSW 2011, Australia

FRANCE   Pergamon Press SARL, 24 rue des Ecoles,
                     75240 Paris, Cedex 05, France

FEDERAL REPUBLIC Pergamon Press GmbH, Hammerweg 6,
OF GERMANY   6242 Kronberg/Taunus, Federal Republic of Germany

Library of Congress Cataloging in Publication Data
Main entry under title:

Nuclear waste management.

    (Pergamon policy studies on energy)
    1. Radioactive waste disposal in the ocean--Congresses.
I. Oceanic Society. II. Georgetown University.
Institute for International and Foreign Trade Law.
III. Center for Law and Social Policy. IV. Series.
TD898.N83   1981   363.7'28   81-11991
ISBN 0-08-027204-5   AACR2

Printed in the United States of America

The Oceanic Society

Executive Offices
Stamford Marine Center
Magee Avenue
Stamford, CT 06902
(203) 327-9786

Christopher Roosevelt, President

Michael J. Herz, Ph. D., Executive Vice President

Thomas C. Jackson, Vice President and
Forum Director

Lloyd M. Wilson, Vice President

Georgetown University
International Law Institute

Don Wallace, Jr., Ph. D., Professor of Law and Director

George Spina, Executive Director

The Center for Law and Social Policy

Clifton Curtis, Attorney

Fig. 1.0   Christopher Roosevelt

1. FOREWORD

Christopher Roosevelt
President, The Oceanic Society

Almost two years ago, a panelist of another Oceanic Society forum on nuclear power and the marine environment, said: "The sea disposal, seabed emplacement option is considered more and more as a viable option at the international level. Both the International Atomic Energy Agency (IAEA) and the Nuclear Energy Agency (NEA) are working on the problem and are developing a framework for conducting -- not banning -- sea disposal. They are planning on more of it. This is coming across to us in the United States. Right now, the Nuclear Regulatory Commission (NRC) and the Department of Energy (DOE) are both seriously looking at this waste management option." Today, almost two years later, the situation has not changed dramatically.

Since that 1978 Oceanic Society Public Policy Forum, President Carter's Interagency Review Group on Nuclear Waste Management has confirmed the interest of this country in considering the "ocean alternative." Through DOE's Subseabed Disposal Program, the federal government is directing a significant flow of tax dollars for the scientific study of this option for disposal of high-level wastes.

Low-level radioactive wastes continue to be dumped in the northeast Atlantic Ocean by several nations. The current trend is for review of the "ocean alternative" as a means of disposal of higher volumes of nuclear waste. Other countries are, after a period of abstention, considering resumption of sea disposal for low-level wastes. One nation alone is scheduled to dispose of more than 5,000 containers of radioactive wastes once its government ratifies the London Dumping Convention. Clearly, the "ocean alternative" merits our attention as a significant element in the nuclear waste management equation.

To quote Dr. Christian Patterman of the West German Federal Ministry of Research and Technology: "The problem of radioactive waste management has become an issue of world-wide interest among politicians, industrialists, scientists and citizens. In many states, the future use or expansion of nuclear energy -- and I might add the use of radioactive materials in science and medicine -- has become dependent on or at least closely interrelated with the ability of effectively operating waste management."

And to quote Dr. David Deese who will be participating in the afternoon session: "United states decision-making in radioactive waste management has been in a state of flux over the last four years. As a now highly politicized and emotional issue, in at least Britain, Japan, Sweden, United States and West Germany it has captured considerable high level management attention. Despite the

1

complexity there are broad scale and far reaching ramifications of
the radioactive waste management issue."

We are here today to focus on the technical, legal, and policy
questions relating to the "ocean alternative." As part of this process, we will examine some of the unique considerations and constraints involved with analysis of ocean dumping for low-level wastes
and sub-seabed emplacement for high-level wastes. We shall also hear
the Hon. Elliot Richardson, United States Ambassador to the Law of
the Sea Negotiations, discuss some of the international implications
of the "ocean alternative." A discussion of United States policy
trends within a global context will close today's forum.

This public policy forum is just one facet of the Oceanic Society's work. A non-profit membership organization, The Oceanic Society is devoted to preservation and wise use of our oceans through
education, research, and conservation programs. Our other projects
range from publications of OCEANS magazine as our membership journal
to marine environmental policy analysis, development of local chap-
ters, and a wide ranging ocean education program.

Through this public policy forum, we are seeking to sponsor
rational and informed decision making in an area marred previously
by emotion and bias. We encourage the audience to participate in
this forum with questions, comments and discussion. Ample opportunity will be provided by each panel for your involvement. We only
ask that you confine your participation to the subject at hand.

The Oceanic Society would like to express our appreciation to
the Georgetown University Law Center (my own law school alma mater)
for the use of this fine Moot Courtroom. In particular, the courtesy
and assistance of Professor Don Wallace and Mr. George Spinna of the
Georgetown Law Center Institute for International and foreign trade
have been most helpful in designing and making arrangements for this
forum. Second, we would like to acknowledge the significant involvement and support of Clifton Curtis, an attorney with the Center for
Law and Social Policy who worked closely with Jane Whitehead and
Thomas C. Jackson of The Oceanic Society in putting this forum together. Cliff will be presenting his paper regarding United States
policy trends this afternoon and Tom will be joining me at the
table here on your right and moderating the panel discussion. We
would like to thank Robert S. Dyer of the Office of Radiation Programs of the United States Environmental Protection Agency for providing us with his paper on a brief history of the sea disposal of
nuclear waste. Copies of that paper are available to members of the
audience. Bob will also be joining us this morning on the panel discussion of low level waste disposal.

Finally, on behalf of The Oceanic Society and the participants of
today's public policy forum, we would like to thank Sidney Shapiro
and the Max and Anna Levinson Foundation for supporting this session.
With this assistance, The Oceanic Society has convened for the first
time a public forum examining the full range of implications raised
by government review of the "ocean alternative" for nuclear waste
disposal.

2. INTRODUCTION

Thomas C. Jackson
Vice President
The Oceanic Society

"If you folks are actually considering 'the ocean alternative' as a way to deal with this nuclear insanity, you are crazy and ought to be considered armed and extremely dangerous and dealt with accordingly."

This citizen's response to notice of our forum reflects a difficulty the Oceanic Society faces in promoting balanced discussion of nuclear waste disposal in the marine environment. It also underscores a continuing need to involve the public in reasoned debate on proposed ocean policies. Some of this individual's anger stems from a sense of surprise the oceans should even be considered for disposal of radioactive wastes.

Yet some nations have moved well beyond consideration of this option and now routinely dispose of low-level nuclear wastes in the North Atlantic. And the United States -- which no longer dumps radioactive materials in our coastal waters -- is seriously studying prospects for disposal of high-level wastes through burial in geological formations of the deep seabed.

This forum was designed to increase public understanding of the ocean option in the complex problem of nuclear waste disposal. By sponsoring this session, the Oceanic Society neither endorsed nor opposed a policy of radioactive disposal in the marine environment. Rather, the Society convened this session to stimulate a rational and informed discussion on the subject and to focus public attention on an issue of great concern to those who care about our oceans.

This proceeding does not contain all of the answers. But it does include critical conservation concerns which must be addressed in drafting ocean dumping policy. This publication goes beyond merely raising questions to include a summary of current scientific knowledge, legal principles and political thought on the "ocean alternative."

Until this forum, discussion of the ocean option had not moved beyond a relatively small circle of scientists, technical experts and public officials. Despite this, federal policy continues to list subseabed disposal as a principal alternative for high level wastes should land sites be deemed scientifically -- or politically -- unsuitable.

Through this publication, the Oceanic Society is working to increase public understanding of this issue. Acting through our offices in Connecticut and California we are building a constituency of citizens and conservationists. With their support, the

Society will continue to speak as a knowledgeable and responsible
voice in the   nuclear waste debate.

Finding an "acceptable" method for disposing of nuclear waste
stands as one of this era's great unresolved questions. Proposals
for resolving this problem have ranged from burial of low-level
wastes at sea to rocketing high-level wastes into outer space.
Currently, the U.S. does neither.

Today, federal policy calls for disposal of low-level wastes on
land although the U.S. Environmental Protection Agency does have
the authority to establish criteria and issue permits for ocean
disposal. Government policy for disposal of high-level wastes
favors subsurface land disposal and continued scientific study of
the "ocean alternative."

At issue here is finding a nuclear waste management policy which
insures radioactive materials will be isolated from the biosphere
until the danger of environmental degradation has passed. How
phases like "isolated from the biosphere" and "danger of environmental degradation" are defined shapes the direction and extent
of a continuing policy debate.

Nuclear wastes were first produced by the U.S. weapons program in
the 1940's. To date, some 66 million gallons of highly radioactive wastes have been generated by the defense program. These
materials are being stored in liquid form at temporary facilities
on three federal installations.

A decade later, commercial sources began to add to the pile of
nuclear waste as atomic power was harnessed to generate electri-
city. Measured in terms of radioactivity, commercial reactors have
generated more overall radioactivity than the weapons program. This
trend is expected to continue as long as new nuclear power plants
go into use across the country. Like military waste, commercial
nuclear waste is stored temporarily, often in the reactor's on site
storage facilities, awaiting a permanent solution.

A "final solution" to the nuclear waste problem is years away.
Depending on the elements involved, radioactive waste can take hundreds, thousands or even millions or years to decay to stable materials. An ultimate answer to this dilemma must isolate these wastes
for periods of time which are almost beyond comprehension in a
country which recently celebrated its bicentennial.

Disposal of radioactive wastes underground in geological formations
of rock or salt is the prime solution currently advanced for high-
level wastes. Salt beds are the most thoroughly researched and
most serious candidate. But the corrosive nature of salt -- and
the threat radioactivity could leach into the groundwater -- has
led some scientists to question this medium and urge further
studies of rock deposits. Granite, basalt and shale are among the
rocks suggested for study.

Scientists say nuclear wastes are potentially dangerous as long as
the material remains highly radioactive. Exposure to sufficiently
high doses of radiation can lead to death. In lower doses --
perhaps even very low doses, studies have linked radiation to

mortal diseases including cancer. It can also cause mutagenic
changes in humans, animals and plants, posing unknown dangers to
future generations. Today, it is difficult to define a "safe exposure" limit for individuals. Some scientists support levels set
for maximum safe exposure set by government agencies. But other
scientists dispute this, arguing "safe exposure limits" should be
significantly reduced.

SOURCES OF NUCLEAR WASTE

Exposure to radiation comes from many sources. Radioactivity
exists in nature and weapons tests have added to the background
exposure level. The two principal types of nuclear waste under
discussion here are: low-level waste from civilian and military
programs and high-level waste from reactors.

Low-level waste includes equipment and materials used in the weapons program, power plant operation, medical activities, and industrial nuclear activities. The U.S. has 15.8 million cubic feet of
low-level commercial waste and 50.8 million cubic feet of defense
waste.

Current disposal means for these wastes, which can remain radioactive for several thousand years, is through burial in shallow
earthen trenches about 20 feet deep. Both the Department of Energy
and commercial firms operate low-level land disposal sites. But
more and more of the commercial sites are being pressured by state
officials to reduce disposal activity and some states have moved
toward banning wastes from other states.

High-level waste comes principally from the highly radioactive fuel
rods used in nuclear power reactors. Once the reactor has used the
available fissionable material, reprocessing can dissolve this
spent fuel in acid, recovering unused uraniun and plutonium for
reuse.

Reprocessing leaves behind highly radioactive liquid wastes containing "fission products" and small levels of uranium, plutonium
and other transuranic elements. Fission products are the smaller
atomic fragments left from the uranium atoms which split inside
the reactor during the chain reaction. Transuranic elements are
heavier atoms formed when other non-fissionable atoms capture neu-
trons inside the reactor.

Most fission products loose their radioactivity more quickly than
transuranics and they account for much of the radioactivity and
heat emitted by the waste during the first few centuries of decay.
Two exceptionally long-lived fission products are iodine-129 which
concentrates in the thyroid, and techentium-99 which concentrates
in the gastrointestinal tract.

The fission products lose most of their radioactivity in the first
700 years, after which time the principal source of radioactivity
comes from man-made transuranic elements. Transuranic elements can
take hundreds of thousands of years more to lose their radioactivity. Plutonium-239, one of the most toxic transuranics, can cause
cancer if inhaled in minute quantities. This element has a half-

life of 24,000 years, which means it takes 24,000 years to lose
half of its radioactivity and then another 24,000 years to lose
half of its remaining activity, etc.

Relatively little high-level wastes have been produced from commercial power plant fuels since so few spent fuel rods have been reprocessed to date. A recent Interagency Review Group report to the
President showed 80,000 cubic feet of high-level commercial waste.
In comparison, military programs have generated almost 120 times
that quantity of high-level wastes.

Since the vast majority of spent fuel rods have not been reprocessed, they can - at least for the time being -- be considered
as part of the high-level waste load. Technically, since spent
fuel contains recoverable reusable radioactive elements, some
people argue it cannot be considered as "waste." Nevertheless,
the fact remains some 2,300 metric tons of spent fuel rods have
been discharged from commercial reactors. In terms of radioactivity, this material contains as much radioactivity as the military's high level waste. These spent fuel rods compose the bulk of
the commercial radioactive waste disposal problem.

For final storage, liquid high-level wastes must be solidified for
stability and safe handling. High-level waste is characterized
by high levels of penetrating radiation, high rates of heat generation and long half-life.

Some scientists say the problems posed by nuclear waste management
are insurmountable and should lead to the end of commercial atomic
generating plants. Other scientists argue with just as much vigor
these difficulties can be overcome and the use of nuclear power
expanded. That issue is beyond this project. Instead, we are
focusing on the "ocean alternative" to insure marine disposal will
never be the option of convenience or expediency in facing the difficult questions posed by the management of nuclear waste.

As part of this effort, this book has been published in cooperation
with Pergamon Press. Clifton Curtis of the Center for Law and Social
Policy, George Spinna of the Georgetown University Law Center Institute for International and Foreign Trade Law, and Sidney Shapiro of
the Max and Anna Levinson Foundation all aided in this effort;
without their assistance and advice the forum could not have suc-
ceeded. Finally, a special thanks to all the Oceanic Society staff
who have joined in the planning and presentation of the forum as
well as preparation of this manuscript for publication.

Fig. 2.0   Forum Participants

Robert Dyer

Fig. 3.1   Robert Dyer

3. SEA DISPOSAL OF NUCLEAR WASTE: A BRIEF HISTORY

Robert S. Dyer
Office of Radiation Programs
U.S. Environmental Protection Agency

ABSTRACT

Past practices and policies for sea disposal of radioactive wastes are examined in this paper. After reviewing the scope of American sea disposal programs between 1946 and 1970, Mr. Dyer turns to a discussion of the concentrations of radioactive wastes at 35 dump sites used by the United States. The U.S. decision to halt sea disposal of low-level radioactive wastes in 1970 and current federal laws are also discussed. International regulations based on the London Dumping Convention and a review of sea disposal practices by other nations are examined during this paper. Mr. Dyer's analysis served as the principal background paper for the Forum.

UNITED STATES PRACTICES

Between 1946 and 1970, the United States dumped more than 60,000 curies of packaged, solidified, low-level nuclear waste into its coastal and offshore waters at depths ranging from 50 feet to more than 15,000 feet. Much of the waste consisted of contaminated laboratory clothing, glassware, experimental animals, and radioactive liquids from laboratory experiments. Some wastes derived from weapons production were also included. Most of this low-level radioactive waste was packaged in 55-gallon drums filled with concrete to ensure that the drums were of sufficient density to sink to the ocean bottom. However, these packages were neither designed nor required to remain intact for sustained periods of time after descent to the sea bottom and it was assumed that all the contents would eventually be released.[1]

The United States conducted sea disposal operations at more than 35 ocean dump sites. The majority of the dump sites were located in the Atlantic Ocean, with the remainder of the sites located in the Pacific, except for two sites in the Gulf of Mexico which received very little use.[2] However, most of the volume as well as radioactivity inventory was deposited in only four of these more than 35 dump sites: two sites in the Atlantic Ocean off the Maryland-Delaware Coast, and two sites in the Pacific Ocean off San Francisco, California, near the Farallon Islands. These four dump sites received more than 90 percent of the radioactive waste packages and 95 percent of the estimated radioactivity dumped. Table 3.1 provides a summary of the U.S. radioactive waste dumping activities at these four major dump sites.[3]

All U.S. sea disposal of nuclear waste was carried out under the
direction of licensing authority of the former Atomic Energy Commission (AEC). In both the Atlantic and Pacific Oceans, more than
90 percent of the radioactive waste was both generated and disposed of
by AEC contractors or defense facilities. Atlantic sites received
wastes from the Brookhaven National Laboratory and Bettis Atomic Power
Laboratory while Pacific sites received materials from the Lawrence
Berkeley Laboratory, Lawrence Livermore Laboratory, and U.S. Naval
Radiological Defense Laboratory. The United States Navy assisted with
the actual transport of wastes to sea until 1959 when private companies assumed the responsibility under AEC license. In 1960, the AEC
placed a moratorium on the issuance of new licenses for sea disposal
of nuclear waste and designated its facilities at Oak Ridge, Tennessee,
and Idaho Falls, Idaho, as interim low-level waste burial sites for
AEC licensees. In 1962, the first permanent commercial land disposal
site for low-level radioactive waste was established at Beatty, Nevada. Because of both increasing public concern with sea disposal and
greater economy of land disposal, the AEC contractor facilities turned
shortly thereafter to the land disposal option."

ENVIRONMENTAL SAMPLING

During the 25 year period that the United States conducted sea disposal of nuclear waste, three surveys were made. In the Pacific dumpsite areas two surveys were carried out under AEC sponsorship: first,
in 1957, by the Scripps Institution of Oceanography; and again, in
1960, by the Pneumodynamics Corporation. The radioanalytical results
of the 1957 survey of water, sediments, and marine organisms did not
disclose "easily detectable amounts of radioactivity even in the immediate vicinity of the oldest official marine disposal area."[5] The
1960 survey also demonstrated that "within experimental error there
was no radioactivity detected that exceeded background levels" in the
water, sediments, or biota.[6] In 1961, the U.S. Coast and Geodetic
Survey investigated the Atlantic 2,800m and 3,800m disposal site areas
under an AEC contract.[7] Analysis of sediments and water from these
sites gave no positive indication of radioactivity releases.

However, as one of the surveys concluded, better techniques and
measurement systems needed to be perfected for more efficient and sensitive monitoring of these dump site areas.[8] It should also be noted
that more than 11,000 underwater photographs were taken during the
1960 and 1961 Pacific and Atlantic dump site surveys, but none of the
more than 50,000 radioactive waste drums that had been reportedly
dumped in these sites were seen. This could suggest that much of the
early survey work was not conducted in the immediate vicinity of the
radioactive waste packages which were actually dumped.

The United States discontinued sea disposal of nuclear waste in 1970,
following the recommendations of the federal Council on Environmental
Quality (CEQ) in a Report to the President. This report recommended
a continued policy of prohibiting ocean disposal of high-level radioactive waste. It further recommended that, "Dumping other radioactive
materials would be prohibited except in a very few cases for which no
practical alternative offers less risk to man and his environment."[9]

These CEQ policy recommendations were further reflected in the AEC
regulations providing standards for protection against radiation,

to wit, "The Commission will not approve any application for a license for disposal of licensed material at sea unless the applicant shows that sea disposal offers less harm to man or the environment than other practical alternative methods of disposal."[10]

Congress codified the CEQ policy recommendations with the passage of Public Law 92-532, The Marine Protection, Research, and Sanctuaries Act of 1972. This law continues the prohibition on sea disposal of any high-level radioactive waste or radiological warfare agent. It further designates the United States Environmental Protection Agency (EPA) as the responsible federal agency for establishing and administering a permit review and evaluation program for the ocean disposal of any waste including radioactive waste not prohibited by law. EPA has not issued any permits for sea disposal of nuclear waste and since 1970, the United States has not carried out any disposal of nuclear waste in the oceans.

Recent Surveys

From 1974 to 1978, EPA's Office of Radiation Programs conducted a series of surveys at the four major U.S. ocean disposal sites. The primary objectives of these surveys were to:
1. determine whether these sites could be surveyed directly using manned and remotely-controlled submersibles,
2. assess the condition of the radioactive waste packages, and
3. determine whether radioactive material was released.
All of the sites were successfully examined with submersibles. The condition of the packages examined ranged from very good with little surface corrosion to very poor with severe hydrostatic implosion. Radioactivity at low concentrations was detected around the packages in both the Atlantic 2,900m disposal site (cesium-137), and the Pacific Farallon Islands disposal sites (plutonium-238, and plutonium-239, 240).[11] [12]

With increasing public concern for waste management practices on land and the need to find permanent disposal sites, the United States is again looking towards the oceans as a possible alternative to land disposal for both low-level and high-level radioactive waste. Many other nations are also now using or considering the future use of this ocean disposal alternative, particularly for their low-level radioactive waste.

INTERNATIONAL PRACTICES

During the early years of nuclear technology development, some isolated instances of sea disposal were carried out by countries such as France and Japan. However, the United Kingdom was the principal user of this option between the years 1950 and 1966. During this period they reportedly disposed of approximately 47,000 curies of alpha and beta-particle emitting wastes.[13] These wastes were dumped in the Northeast Atlantic near the Bay of Biscay and also in the Hurd Deep about twenty miles north of Guernsey Island in the Channel Islands. Most of this waste was packaged in 55 gallon drums weighted with concrete or bitumen.

In 1967, a significant event occurred in the evolution of international supervision of nuclear waste disposal in the ocean. The Nuclear Energy Agency (NEA) of the Organization for Economic Cooperation and

Development (OECD) agreed to accept supervisory responsibilities for
sea disposal of low-level nuclear waste by NEA Member countries. The
primary objectives of OECD/NEA were "to develop, at the international
level, a safe and economic method for ocean disposal and to demonstrate this by a joint experimental disposal operation involving
several Member countries."[14] Five countries participated in this
first sea disposal operation in 1967 under international supervision
(Belgium, France, Federal Republic of Germany, Netherlands, and the
United Kingdom).

During the next internationally-supervised sea disposal operation in
1969, Italy, Sweden, and Switzerland also participated, while the
Federal Republic of Germany abstained. Since 1971, however, only
Belgium, Netherlands, Switzerland and the United Kingdom have used
the sea disposal option. Beginning in 1967, all internationally-
supervised sea disposal has occurred in the Northeast Atlantic Ocean
at depths in excess of 11,000 feet. Table 3.2 presents an inventory,
by year, of the mass of material dumped (which includes the packaging
materials) as well as the estimated alpha and beta-gamma activity of
the wastes at the time of packaging.[15, 16] Examples of radioactive
isotopes that were dumped in the ocean include radium-226, and plu-
tonium-239, (alpha-emitters), strontium-90 (beta-emitter), and cobalt-
60, zinc-65, and cesium-137 (beta-gamma emitters). It is interesting
to note in Table 3.2 that with the exception of the year 1974, the
inventory of alpha-emitters disposed into the ocean has increased
each year. The inventory of beta-gamma emitters has also shown a
substantial increase in later years compared with the first NEA-super-
vised sea disposal operations.

Between 1967 and 1979, more than 500,000 curies of alpha plus beta-
gamma-emitting radioactive wastes have been disposed into the Northeast Atlantic Ocean.

Another important milestone in the international sea disposal of
nuclear waste occurred in London in 1972 with the development of the
Convention on the Prevention of Marine Pollution by Dumping of Wastes
and Other Matter in the Oceans. Annex I of this Convention continued
the prohibition on sea disposal of high-level radioactive waste and
called upon the International Atomic Energy Agency (IAEA) to define
this material on public health, biological, or other grounds. The
definition would become a mandatory limit for signatory countries
(Contracting Parties) to this Convention. Annex II of the Convention
lists material requiring special care when considered for sea disposal.
Radioactive materials, not otherwise prohibited, are included in this
list.

The IAEA was also given the responsibility to develop recommendations
which should be used by Contracting Parties in issuing permits for
sea disposal of nuclear waste. As of January 1980, 41 countries, including the United States, have ratified or acceded to the London
Convention. Of those countries which have carried out sea disposal
under supervision of NEA/OECD, only Belgium and Italy have yet to
ratify the Convention.

Other recent documents providing for further international controls on
sea disposal of nuclear wastes include: the IAEA Definition and
Recommendations developed pursuant to Annexes I and II of the London
Convention[17], and a Consultation and Surveillance Mechanism developed
by the OECD clearly specifying the requirements that any NEA member

country must follow when evaluating the sea disposal option[18]. Note-
worthy aspects of the IAEA Definition and Recommendations include:

    1. a numerical definition of high-level radioactive waste prohibi-
ted from sea disposal;
    2. a philosophy of isolation and containment, rather than dilution
and dispersion, of radioactive waste after it reaches the seabed; and
    3. a minimum acceptable disposal depth of 4,000m. Of particular
interest in the OECD Consultation and Surveillance Mechanism is the
requirement for prior notification of proposed sea disposal action,
as well as site suitability assessment and review of existing and past-
used radioactive waste dumpsites.

Since 1967, almost all international sea disposal of nuclear waste has
been conducted under international supervision and controls. Although
the administrative mechanisms are actively being developed, the scientific predictive capability for the movement of radioactivity released
from the drums into such deep-sea areas as the Northeast Atlantic disposal site is clearly lacking. To date, no oceanographic multi-disci-
pline survey of any of the Northeast Atlantic dumpsite areas has been
conducted.

We must learn from our past sea disposal experiences. The immediacy
of the conclusions made 18 years ago by a working group of the Committee on Oceanography of the National Academy of Sciences - National
Research Council is still with us today:

    "There is no evidence that this disposal practice has resulted in
    any inimical effect upon the environment; but neither is there
    evidence that harmful effects cannot eventually result from it.
    The concern here is not with any magnitudes of disposal already
    undertaken, but rather with understanding the implications of
    the continuing and increasing use of the oceans as a receptacle
    for disposal. History is replete with cases in which unrestricted
    pollution of various kinds, rapidly developing from innocuous
    beginnings, has subtly damaged or destroyed resources before
    understanding and controls could be developed."[19]

The controls are being put in place, and scientific understanding of
the consequences of this marine disposal activity must be rapidly ob-
tained while we are still at the "innocuous beginnings."

Robert S. Dyer

Fig. 3. 0

SUMMARY OF
NEA-SUPERVISED SEA DISPOSAL OPERATIONS
FOR NUCLEAR WASTE
IN THE
NORTHEAST ATLANTIC OCEAN

R.S. Dyer

YEAR	DUMPED WEIGHT (tonnes)	APPROXIMATE ACTIVITY	
		ALPHA (Ci) (actinides)	BETA-GAMMA (Ci) (incl. ^{3}H)
1967	10,840	250	7,600
1969	9,180	500	22,000
1971	3,970	630	11,200
1972	4,130	680	21,600
1973	4,350	740	12,600
1974	2,270	420	100,000 *
1975	4,460	780	60,500 **
1976	6,770	880	53,500 ***
1977	5,605	958	76,450 ****
1978	~8,000	1,100	79,000 *****
1979	~5,400	1,400	83,000 ******

* TRITIUM ALMOST EXCLUSIVELY
** INCLUDING ABOUT 30,000 Ci OF TRITIUM
*** INCLUDING ABOUT 21,000 Ci OF TRITIUM
**** INCLUDING ABOUT 32,000 Ci OF TRITIUM
***** INCLUDING ABOUT 36,000 Ci OF TRITIUM
****** INCLUDING ABOUT 42,000 Ci OF TRITIUM

BETWEEN 1967-1976 THE DUMP SITE USED WAS A CIRCLE OF 70 NAUTICAL MILES DIAMETER CENTERED ON THE POINT 46° 15' NORTH BY 17° 25' WEST. THE DUMP SITE HAS AN AVERAGE DEPTH OF 4.5 km AND IS APPROXIMATELY 900 km SOUTHWEST OF LAND'S END, ENGLAND.

SINCE 1977 THE DUMPING SITE USED IN THE NORTHEAST ATLANTIC HAS BEEN DEFINED AS A RECTANGLE BOUNDED BY THE COORDINATES 16° W TO 17° 30' W AND 10 NAUTICAL MILES NORTH AND SOUTH OF 46° NORTH LATITUDE

REFERENCES

1. Disposal of Low-Level Radioactive Waste into Pacific Coastal Waters, National Academy of Sciences - National Research Council Publication No. 985, Washington, D.C., (1962).

2. Smith, David D. and Robert P. Brown, Ocean Disposal of Barge-Delivered Liquid and Solid Wastes from U.S. Coastal Cities, prepared for U.S. Washington, D.C., (1971).

3. Dyer, R.S., "Environmental Surveys of Two Deepsea Radioactive Waste Disposal Sites Using Submersibles", Management of Radioactive Wastes from Nuclear Fuel Cycle, Vol. II, IAEA, Vienna, (1976).

4. Press Release, U.S. Atomic Energy Commission, San Francisco Operations Office, SAN No. 298,(September 30, 1962).

5. Faughn, J.L. et al., University of California Project Report: Radiological Survey of the California Disposal Areas, prepared for the U.S. Atomic Energy Commission and the Office of Naval Research, (1957).

6. Pneumodynamics Corporation, Survey of Radioactive Waste Disposal Sites, TID-13665, U.S. Atomic Energy Commission, (1961).

7. Jones, Edmund L., Special Report, Waste Disposal Program Project No. 10,000-827 of U.S. Coast Guard, prepared for U.S. Atomic Energy Commission, (1961).

8. Op. cit., Faughn, J.L. et al.

9. Ocean Dumping: A National Policy, Report to the President prepared by the Council on Environmental Quality, Washington, D.C., (1970).

10. Conditions and Limitations on the General License Provisions of 10 CFR 150.20, Rules and Regulations, U.S. Atomic Energy Commission, Washington, D.C., (1972).

11. Radiological Contamination of the Oceans, Oversight Hearings before the Subcommittee on Energy and the Environment of the Committee on Interior and Insular Affairs, U.S. House of Representatives, Serial No. 94-69, Washington, D.C., (1976).

12. Ocean Dumping and Pollution, Hearings before the Subcommittee on Oceanography and the Subcommittee on Fisheries and Wildlife Conservation and the Environment of the Committee on Merchant Marine and Fisheries, U.S. House of Representatives, Serial No. 95-42, Washington, D.C. (1978).

13. Radioactivity in the Marine Environment, prepared by the Committee on Oceanography, National Research Council, National Academy of Sciences, Washington, D.C., (1971).

14. Radioactive Waste Disposal Operation Into the Atlantic - 1967, European Nuclear Energy Agency, Organization for Economic Cooperation and Development, Paris, (1968).

15. Assessing Potential Ocean Pollutants, A Report to the Ocean Affairs Board of the Commission on Natural Resources, National Re-

search Council - National Academy of Sciences, Washington, D.C. (1975).

16. Organization for Economic Cooperation and Development, Activity Reports Number 9-13 of the European Nuclear Energy Agency (1967-1971), and Activity Reports Number 1-7 of the Nuclear Energy Agency (1972-1978), OECD/NEA, Paris.

17. The IAEA Revised Definition and Recommendations of 1978 Concerning Radioactive Wastes and Other Radioactive Matter Referred to in Annexes I and II to the Convention, IAEA INFCIRC/205/Add. 1/Rev. 1, Vienna, (1978).

18. Decisions of the Council Establishing a Multilateral Consultation and Surveillance Mechanism for Sea Dumping of Radioactive Wastes, OECD, C(77)115 (Final), Paris, (1977).

19. Op. cit., Disposal of Low-Level Radioactive Waste into Pacific Coastal Waters.

4. SEA DISPOSAL OF LOW-LEVEL NUCLEAR WASTES

The concept of dose -- rather than volume of radioactive content --
as a limiting factor in disposal of low-level nuclear waste in the
deep ocean is reviewed by W.L. Templeton in the first section of
this Chapter. After reviewing past and present practice, the basis
of revised definitions and recommendations developed by the International Atomic Energy Agency (IAEA) are examined. The examination
focuses on the oceanographic and radiological basis of the definitions and recommendations before discussing the implications of the
IAEA position. Pathways, modes of exposure, intake-occupancy rates,
and release rate limits are covered during the paper.

After reviewing the three principle strategies for disposal of
hazardous wastes, Dr. Kilho Park emphasizes the need for additional
field work to verify current scientific models. During his forum
comments, Dr. Park reviews current studies and points to the need for
additional investigation of deep-sea ecology. Noting the wide range
of species found on the deep seabed, Dr. Park characterizes modeling
of these species as a "formidable task" and calls for construction
of a 6,000 meter deep-sea submersible for use in future research.

Four years of Environmental Protection Agency (EPA) field work investigating release of radioactivity from low-level wastes dumped
by the United States are summarized by Robert Dyer in the next section of this chapter. Mr. Dyer reviews data developed during 1974,
1975, 1976, 1977 and 1978 through use of submersible equipment. He
lists current concerns and includes 15 photographs from EPA research
projects.

This chapter closes with a brief question and answer discussion
which centers on wastes placed in the Farrallon Island site near
San Francisco, California. (Editor's note).

THE BASIS OF THE REVISED INTERNATIONAL ATOMIC ENERGY AGENCY DEFINITION AS RELATED TO THE DUMPING OF LOW-LEVEL RADIO-ACTIVE WASTES IN THE DEEP OCEAN

W.L. Templeton
Ecological Sciences Department
Pacific Northwest Laboratory
Batelle Memorial Institute

The oceans of the world have been used since time immemorial as a repository for man's wastes. The majority of these wastes are natural
products which, when degraded, enter the natural biochemical cycles
of the seas with little or no impact upon the ecosystem. However,
during the last several decades, man has not only increased his use
of the oceans as a waste depository but has introduced increasing
quantities of industiral wastes containing manufactured inorganic and

17

William Templeton

Fig. 4.0 William Templeton

organic chemicals. Many of these are toxic and in certain cases result in tragic consequences to man.

In many cases these problems arise because of a lack of knowledge of
the basic properties of the materials and the processes in the marine
environment. Without controls, some of these systems become over-
loaded with material so that natural processes are depressed; e.g.,
reduced oxygen content and the ecosystem operates with reduced effi-
ciency. In other cases the amounts of inorganic and organic toxic
materials discharged are so large as to result in debilitation or
death of the marine biotic resources. Sometimes the toxic materials
are transferred by the food web resulting in toxic materials being
incorporated into the diets of higher trophic organisms, including
man.

In order to control coastal discharges or ocean dumping, it is necessary to determine a release rate. This can only come from a knowledge of the composition and chemical form of the source materials;
the distribution and bioavailability of these materials in the ocean
ecosystem; the degree and rates of bioaccumulation; and the actual
or potential use of the ocean resources. With this information release rates within acceptable limits for man and the ecosystem can
then be determined. Today, probably the only situations which apply
this approach are the controlled disposal of radioactive wastes.

The most important criteria for assessing the impact of present day
waste management practices, or a comparison of nuclear waste management alternatives, is that of radiation dose. The gross weight of
concrete and metal is not relevant except from an operational standpoint. The number of curies -- whether it be tens, hundreds, or
thousands -- is pertinent only as regards the derivation of the resultant radiation dose to man and the biota in the environment.

One cannot directly equate the impact of a curie of plutonium or
other long lived actinides with a curie of potassium-40 in the environment without a determination of the resulting doses since each and
every radionuclide is unique in decay rate; in rate of transport and
movement; in degree of bioaccumulation; in degree of receptor exposure; and in the effect per unit ionization at or in the receptor.
We know, for example, the oceans contain hundreds of billions of
curies of natural radionuclides (some of which are starting materials
for the nucelar fuel cycle) including potassium-40, rubidium-87,
uranium-238, radium-226, carbon-14, uranium-235 and thorium-232; hundreds of megacuries of artificial radionuclides produced by weapon
tests; and a few megacuries from the controlled disposal (solid and
liquid) from defense and commercial nuclear operations. We have
been able to predict and validate the resultant radiation exposures
from each of these sources that many and biota in the environment
have been exposed to.

In this paper I will discuss an extension of that knowledge to a
radiological assessment of the controlled dumping of packaged low
level radioactive wastes to the deep ocean seabed.

Past and Present Practice

In the 1950's and 1960's many countries used the oceans for dumping

of packaged low-level radioactive wastes. Between 1946 and 1970 the
United States dumped more than 60,000 curies of packaged low level
radioactive wastes into the offshore waters of the Atlantic and Pacific Oceans. This does not include the reactor pressure vessel of
the nuclear submarine Seawolf with an estimated induced activity of
33,000 curies. In 1972 the United States of America banned the transport, for dumping, of high-level radioactive wastes. Although the
dumping of low-level radioactive wastes are allowed by law, since
that date no permits have been issued for dumping of this type of
materials.

At the same time there were intensive efforts internationally to
reach agreement on ocean dumping of all pollutants. The outcome was
the Convention on the Prevention of Marine Pollution by Dumping
Wastes and Other Matter in the Oceans (the London Dumping Convention
of 1972).[1] The International Atomic Energy Agency (IAEA) was
charged with the task of defining radioactive wastes unsuitable for
dumping at sea and providing recommendations to ensure that any dumping of radioactive material into the oceans involves no unacceptable
degree of hazard to humans and their environment. In 1974 the IAEA
made a provisional definition along with a recommended basis for
issuing special permits.[2]

The provisional Definition and Recommendations of 1974 stated that
high-level radioactive wastes or other high-level radioactive matter
unsuitable for dumping means any material with a concentration in
curies per unit-gross mass (in tonnes) exceeding:

(a)   10 Ci/t for alpha-active waste of half life greater than
      50 years (In the case of ^{226}Ra, not more than 100 Ci/yr may
      be dumped at any one site);

(b) 10^3Ci/t for beta-gamma-active waste (excluding tritium) but
      the limit for ^{90}Sr plus ^{137}Cs is 10^2 Ci/t; and

(c) 10^6 Ci/t for tritium.

The definition is based on an assumed upper limit to the dumping rate
of 100,000 tonnes per year at any one site and averaged over a gross
mass not exceeding 100 tons.

Since 1967, European dumping operations have been organized and conducted by the Nuclear Energy Agency (NEA) of the Organization for
Economic Cooperation and Development (OECD). Up until 1970 the European nations dumped a total of about 8.3×10^3 curies of alpha active
material, 2.6×10^5 curies of beta-gamma active material, and 2.6×10^5 curies of tritium, with a total gross weight of 65,000 metric
tons in the Northeast Atlantic.

Basis of the Revised IAEA Definition and Recommendations[3]

The provisional definition and recommendations were actively reviewed
by the IAEA during 1976-1978. Three major aspects were reviewed:
the oceanographic basis[4], the radiological basis[5], and the implications for the Definition and Recommendations.

THE OCEANOGRAPHIC BASIS [4]

Assessment of permissible dumping rates of radionuclides to the oceans must include the calculation of the concentration throughout oceanic basins resulting from localized sources. However, our under-standing of the processes occurring in the deep oceans is insuffi-cient to permit the construction of a single comprehensive model of the movement of radionuclides. The original oceanographic model used for the provisional definition was inapplicable for long-lived iso-topes in finite sized basins. It was considered that the model by Shepherd which includes [6] advection in a finite ocean meets some of the objections raised about the original model used and allows esti-mates to be made of the entire concentration field, although it only approximates the actual oceanographic processes resulting in the dispersion of radioactive nuclides.

The Shepherd model calculates the equilibrium concentration which would be reached from a continuous release of activity maintained in-definitely into the water near the ocean bottom (less than 4,000 m). The model ocean is of finite size and has a horizontal (but no verti-cal) circulation and three dimensional diffusion. Obviously, this is an idealization but it is adequate for defining large-scale long-term concentrations. However, since little is known about the circu-lation of the deep ocean water, poor vertical mixing cannot be assumed for the isolation of radionuclides. Even if this were so, slow vertical mixing could be short-circuited by direct biological transport. It was concluded then[4], that one should not assume any isolation of the surface waters when estimating the dose to man, and that one should calculate not only the long-term average concen-tration in the bottom water for the appropriate part of the ocean basin but also the appropriate maximum concentrations arising from short-term events.

The model only considers the large-scale average distribution of various oceanographic parameters and does not describe short-term processes, either on the large or small scale such as deep vertical upwelling, effects of large-scale topographic features, or strong convective currents. Since deep vertical upwelling, that is a direct transfer of deep bottom water to the surface, is not explicitly in the model basic due to our sparse understanding of the rates of ver-tical diffusion and advection, one cannot assume that disposal of wastes in deep waters provides any isolation from the surface waters when an assessment of the dose to critical groups is being made.

RADIOLOGICAL ASSESSMENT [5]

Oceanographic aspects. With the oceanographic model as a basis, cal-culations of the concentrations of radionuclides in water for the dose assessment included both the long-term concentration in the water for the appropriate part of the ocean basin and the appropriate maximum concentrations arising from short-term events. In both cases these are bottom water concentrations which imply that these levels would be acceptable to surface waters and therefore make it unnecessary to distinguish between hypothetical consumption of deep-sea organisms and more realistic consumption of upper-layer organisms.

Since it is difficult to foresee the time scale over which releases

of radioactive waste may continue, the calculations have assumed that
releases continue for 40,000 years which is approximately the mean
lifetime of plutonium-239. The release rates limits derived are
therefore such that concentrations in the marine environment of long-
lived radionuclides will increase very slowly over several thousand
years towards their limiting values. This is clearly very conservative, however, it does allow waste dumping operations to be reduced
or stopped at any time without exceeding the limiting values. For
example, if the dumping of [239]Pu is continued at the calculated release rate limit, the concentrations of [239]Pu in the ocean will slowly build up approaching the International Commission on Radiological
Protection derived concentration after 40,000 years. If the practice
ceases after 4,000 years then only 10% of the International Commission on Radiological Protection derived limit will have been reached.
For shorter periods of time the oceanographic model suggests the release rate limits might be controlled by short-term processes of advection and upwelling. In order that unrealistic release limits for
very short-lived radionuclides are not estimated it was assumed that
the containment time on the sea-bed was ten years and that three
years decay occurred between the release point and consumption exposure.

Because of a lack of information on the role of sediments in reducing water concentrations, the calculations ignored sorption on sediments. This obviously overestimates water concentrations, and
means that release rate limits for pathways that do not involve
sediments should be conservative. However, for the radiological
assessments of the dose to man or organisms the concentration on the
sediment was calculated assuming it is in equilibrium with the bottom water already calculated. This clearly overestimates the concentration on sediments if there is significant partitioning between
water and sediment, since it ignores the reduction in overall concentration arising from the sorptive capacity of sediments themselves.

Assessment of pathways. The assessment quantified the parameters involved in a number of representative pathways by which many might become exposed to radioactivity after its release on the ocean bottom.
The pathways chosen include some of which are known to exist and
some which may be important in the future (Table 4.1). For all the
possible pathways which were identified the conservative approach was
taken. For example, a pathway in the future may include systematic
fishing at a depth of 4,000 meters, while the deepest presently
known is 2,000 meters. We have no detailed information on the concentration factors for cephalopods or deep-living fish, and for the
present calculation it was assumed that these would be sufficiently
similar to those for surface organisms.

The pathways selected are generalized representatives and the consumption parameters selected are sufficiently general to include
critical groups in all areas of the world. Where individuals are
likely to be members of only a single critical group, the pathways
were evaluated independently. Where they might be members of more
than one critical group; e.g., shore fishermen and beach dwellers,
the limits have been reduced accordingly.

Five individual pathways involving consumption of sea food were considered. These are not intended to represent any particular species
but are examples of general pathways. Consumption rates were as-

TABLE 4.1 Pathways, Modes of Exposure, Intake/Occupancy Rates

Pathway	Mode of Exposure	Intake/ Occupancy Rates
Fish consumption	Ingestion	600 g/day
Crustacea consumption	Ingestion	100 g/day
Mollusc consumption	Ingestion	100 g/day
Seaweed consumption	Ingestion	300 g/day
Plankton consumption	Ingestion	30 g/day
Desalinated water consumption	Ingestion	2000 g/day
Sea salt consumption	Ingestion	3 g/day
Suspension of sediments	Inhalation	Continuous
Evaporation from seawater	Inhalation	Continuous
Swimming	External irradiation	300 hr/yr
Exposure from shore sediments	External irradiation	1000 hr/yr
Exposure from fishermen's gear	External irradiation	300 hr/yr

sumed to be sufficiently large, in a global context, that for each
pathway it would be unlikely that members of one critical consumption group would also be members of another critical consumption
group.

Four pathways leading to exposure of beach dwellers were considered.
Since some individuals would be likely to be exposed to all pathways the derived limits were reduced accordingly. Three miscellane-
ous pathways were also considered and were combined for convenience.

The IAEA Radiological Assessment conducted the calculations for
the pathways for radionuclides which were felt likely to occur in
wastes liable to be dumped into the ocean. The list included fission
products, actinides activation products, and natural radionuclides.

In an equilibrium situation, or in a situation where the concentrations in the water change slowly compared with the biological turn-
over rates of the radionuclides in the biota, the empirical relationship between organisms and water or the concentration ratio was
used.[5] For the purposes of calculation a similar parameter was introduced in the non-ingestion pathways.

Release rate limits

The output from these calculations for both single site and a finite
ocean volume provides the critical pathway for each radionuclide and
is giving rise to the lowest release rate limit. When pathways have
been combined under one critical group; i.e., beach dwellers, the
critical pathway is that which individually would have the lowest
limit. As an example of the output for a finite ocean volume
$(10^{17}m^3)$ the release rate limits for the forty most restrictive
radionuclides are given in Table 4.2.

To meet the present definition under the London Dumping Convention
the radionuclides were initially grouped according to practical considerations and calculated release rate limits. In some cases radionuclides do not appear in the group to which it would seem that they
belong. This is because of known factors not included in the calculations or practical considerations such as the very low predicted
quantities that will occur. The calculated release rate limits for
these groups are given in orders of magnitude based on the more restrictive members of the group.

TABLE 4.2 Release Rate Limits in Ascending Order
for a Finite Ocean Volume of $10^{17}m^3$

Limit Curies/Year	Nuclide	Critical Group
2.8×10^3	Thorium-229	Beach dwellers
6.8×10^3	Iodine-129	Seaweed eaters
1.1×10^4	Radium-226	Fish eaters
1.4×10^4	Thorium-232	Fish eaters
1.6×10^4	Thorium-230	Fish eaters
$5.7 \times 10^{4(a)}$	Neptunium-237	Seaweed eaters
5.8×10^4	Tin-126	Beach dwellers

Limit Curies/Year	Nuclide	Critical Group
5.9×10^4	Technetium-99	Seaweed eaters
8.6×10^4 (a)	Curium-245	Seaweed eaters
8.7×10^4	Plutonium-242	Seaweed eaters
9.2×10^4	Plutonium-239	Seaweed eaters
1.2×10^5	Americium-243	Seaweed eaters
1.4×10^5 (a)	Curium-246	Seaweed eaters
2.6×10^5	Lead-210	Plankton eaters
3.0×10^5	Plutonium-240	Seaweed eaters
5.9×10^5 (a)	Californium-251	Seaweed eaters
6.1×10^5	Carbon-14	Fish eaters
7.3×10^5	Americium-241	Seaweed eaters
1.1×10^6	Uranium-238	Seaweed eaters
1.5×10^6	Americium-242	Seaweed eaters
3.7×10^6	Nickel-59	Fish eaters
3.9×10^6	Zirconium-93	Beach dwellers
3.9×10^6 (a)	Curium-243	Seaweed eaters
4.4×10^6	Plutonium-238	Seaweed eaters
6.8×10^6	Uranium-235	Seaweed eaters
7.6×10^6 (a)	Curium 244	Seaweed eaters
7.8×10^6	Uranium-234	Seaweed eaters
7.8×10^6	Uranium-233	Seaweed eaters
9.1×10^6	Selenium-79	Seaweed eaters
1.2×10^7	Europium-154	Beach dwellers
1.3×10^7	Cobalt-60	Beach dwellers
1.5×10^7	Europium-152	Beach dwellers
2.0×10^7	Cesium-135	Fish eaters
2.3×10^7	Nickel-63	Fish eaters
2.3×10^7 (a)	Palladium-107	Seaweed eaters
4.0×10^7 (a)	Californium-252	Seaweed eaters
6.6×10^7	Strontium-90	Seaweed eaters
1.4×10^8	Antimony-125	Beach dwellers
1.2×10^8	Silver-110	Mollusc eaters
2.2×10^8	Cesium-137	Fish eaters

For administrative convenience and analytical simplicity, Groups A and B were combined to give three groupings according to the basic properties of decay type and half-life, as shown in Table 4.3.

The single site release rate is more restrictive for short-lived radionuclides, and partitioning of wastes between sites can increase the overall limit for the basin as a whole. For long-lived radio- nuclides, the long term finite ocean basin release rate is more re- strictive and partitioning of wastes between sites does not affect the limit for the basin as a whole. However the input of all radio- nuclides into the basin from all sources, including those from other than dumping of radioactive wastes, must be included in any defini- tive assessment of a release rate limit.

TABLE 4.3

| Group | Release Rate Limits (Ci/yr) | |
	Single Site	Finite Ocean Volume (10^{17}m3)
Alpha-emitters, but limited to 10^4 Ci/yr for ^{226}Ra and supported ^{210}Po	10^5	10^5
Beta-gamma-emitters with half-lives of at least 0.5 yr (excluding tritium) and Beta-gamma emitters of unknown half-lives	10^7	10^8
Tritium and Beta-gamma emitters with half-lives of less than 0.5 years	10^{11}	10^{12}

In all cases, the release rate limits derived correspond directly, given the pathways and parameters used, to the International Commission on Radiological Protection (ICRP) dose limits for individual members of the public. The philosophy underlying this procedure and the use of critical groups is described in publications of ICRP. The annual limit for the effective dose equivalent to individual members of the public applies to the average of this quantity in the "critical group;" namely, the group representing the most exposed individuals. If the critical groups are hypothetical and maximizing assumptions are made in their selection, the ICRP maintains the value of 500 mrem for the annual limit. If, however, real critical groups are identified and realistic models are used to assess the annual effective dose equivalent, the ICRP recommends a limit of 100 mrem in a year for exposures of continuous natures expected year after year. It should be stressed that ICRP dose limits provide a lower boundary of an unacceptable range of values. Values above the ICRP limits are to be avoided while values up to the limit are not automatically permitted, however the values permitted must be justified by assessing the net benefits, considering radiological consequences, and alternative procedures. It is anticipated that optimization procedures would usually result in radiation doses lower than the limits.[8]

On the other hand the ICRP dose limits are not threshold values above which undesirable effects begin to appear, but represent dose values corresponding to individual risks approaching unacceptable levels. The maximum permissible annual intakes (MPAI) corresponding to those dose limits were taken from the IAEA Basic Safety Standards.[9] However, the present model calculations need to be recalculated to take account of the effective dose concept of ICRP[8] and the derived annual limits of intake (ALI). The changes do not affect the model nor the definition significantly. The ALI for plutonium[239] is more restrictive than the MPAI by a factor of two or three, while the ALI's for cobalt[60] and radium[226] are in the opposite direction.

In the provisional Definition and Recommendations of 1974, two explicit safety factors of 10^2 were applied to allow for more than one dumping site and to allow for parameters less favorable than those assumed in the assessment. In the proposed revised Definition and Recommendations explicit account has been taken to account for multiple sites in a finite ocean volume and possible extreme events in ocean areas. It is not appropriate, then, to apply additional safety factors for the same reasons to the present assessment. The numerical values depend on the particular radionuclide and set of circumstances and can neither be determined precisely nor be guaranteed; however, it is considered that the release rates given are the best possible estimates which can be made for them at the present time.

An assessment of the potential effects on the biota of the marine ecosystem was conducted and it was concluded that radiation doses arising as a result of releases within the limits of the Definition are not expected to lead to significant adverse effects to populations as a whole.

The technical basis for the present radiological assessment is on release rate limits and not on dumping rates. However, to meet the present requirements of the London Convention it is necessary to express the Definition in terms of a concentration for a single site and an assumed upper limit on mass dumping rate at a single site of 100,000 tonnes/year with the added proviso of release rate limits for a finite ocean volume of $10^{17}m^3$.[3] This results in concentration limits of:

1. 1 Ci/t for alpha-emitters but limited to 10^1 Ci/tonnes for ^{226}Ra and supported by ^{210}Po;

2. 10^2 Ci/tonnes for beta-gamma-emitters with half-lives of at least 0.5 years (excluding tritium) and mixtures of beta-gamma-emitters of unknown half-lives;

3. 10^6 Ci/tonnes for tritium and beta-gamma-emitters with half-lives less than less than 0.5 years.

SUMMARY

The IAEA oceanographic/radiological model was developed as a tool to define that quantity of radioactive waste which would, on the basis of a limited data base, be unsuitable for dumping, i.e., it set an upper limit for single sites and ocean basing. The present model is inherently more restrictive than was used in the previous assessment. The model assumes that dumping continues for a long period of time, i.e., 40,000 years and hence if the dumping of long lived radionuclides at the derived annual release rate limits is continued then the concentration will build up slowly approaching the ICRP derived concentration after 40,000 years. If the practice ceases after, say, 4,000 years, only 10 percent of the ICRP derived limit will have been reached. This is clearly a very conservative approach.

It should be stressed that the ICRP dose limits provide a lower boundary of an unacceptable range of values. Values above the ICRP limits are specifically to be avoided, while values up to the limit are not automatically permitted. These limits should be considered as constraints for optimization procedures, which usually would result in radiation doses much lower than the dose limits. On the

other hand, the dose limits are not thresholds above which undesirable effects begin to appear, but represent dose values corresponding
to individual risks approaching unacceptable levels.

There are, of course, clear needs for additional scientific investigations into the physical, geochemical, and biological processes controlling the behavior of radioactive materials in the deep oceans;
for radiological surveillance programs designed to provide baseline
data to assist in verifying dose assessment models and hypothesis
testing; and in detecting any short or long-term changes or trends in
the marine environment resulting from present or future dumping operations. The research component for the development of refined site
specific models should be given the highest priority since the appropriate radiological surveillance programs can only be determined
when the relevant results of scientific investigations become available.

The deep oceans are a relatively unexplored portion of our oceans
mainly due to the high expense - both in scientific manpower and
funds - and it is hoped that the required scientific investigations
and assessment for dumping or emplacement can be conducted on a basis
of international cooperation even though some nations do not wish to
avail themselves of this waste management option. There are, of
course, national and international political and legal considerations
which will be discussed later in this forum. It will be unproductive
should these become commingled with the research and scientific considerations.

REFERENCES

1. International Atomic Energy Agency. Convention on the Prevention of Marine Pollution by Dumping of Wastes and Other Matter. INFCIRC/ 205. IAEA Vienna (1974).

2. International Atomic Energy Agency. Convention on the Prevention of Marine Pollution by Dumping of Wastes and Other Matter. The Definition Required by Annex 1, paragraph 6, to the Convention, and the Recommendation Required by Annex II, section D. INFCIRC/205/Add 1. IAEA Vienna (1975).

3. International Atomic Energy Agency. Convention on the Prevention of Marine Pollution by Dumping Wastes and Other Matter. The Definition Required by Annex 1, paragraph 6, to the Convention, and the Recommendation Required by Annex II, section D. INFCIRC/205/Add 1/ Rev 1. IAEA Vienna (1978).

4. International Atomic Energy Agency. The Oceanographic Basis of IAEA Revised Definition and Recommendations Concerning High-Level Radioactive Waste Unsuitable for Dumping at Sea. Tech. Doc-210. IAEA Vienna (1978).

5. International Atomic Energy Agency. The Radiological Basis of the IAEA Revised Definition and Recommendations Concerning High-Level Radioactive Waste Unsinkable for Dumping at Sea. Tech. Doc-211. IAEA Vienna (1978).

6. Shepherd, J.G. A Simple Model for Dispersion of Radioactive Wastes Dumped on the Deep Seabed. Fisheries Research Technical Report No. 29. Ministry of Agriculture, Food and Fisheries U.K. (1976).

7. International Commission on Radiological Protection. "Principles of Environmental Monitoring Related to the Handling of Radioactive Materials," A Report by Committee 4, ICRP Publication 7 Pergamon Press (1965).

8. International Commission on Radiological Protection. Recommendations on the International Commission on Radiological Protection, Publication 26, Pergamon Press (1977).

9. International Atomic Energy Agency, Safety Series No.9: "Basic Safety Standards for Radiation Protection," STI/PUB/147, IAEA Vienna (1967).

POTENTIAL PRIORITIES FOR SCIENTIFIC STUDIES OF DEEP-SEA LIFE

Dr. Kilho Park
Ocean Dumping and Monitoring Division
National Oceanic and Atmospheric Administration

INTRODUCTION

There are essentially three waste management strategies on Earth.
They are: destruction of harmfulness; isolation and containment of
toxic wastes; and dilution and dispersion. On radioactive wastes
the first strategy of destruction of harmfulness does not work most
of the time. Radioactive decay into harmless elements can be very
time-consuming. Consequently, the present state of the waste management strategies is inclined toward isolation and containment for
high-level radioactive wastes and dilution and dispersion for low-
level wastes. Of course, the isolation and containment strategy can
be used for low-level wastes.

As Mr. Templeton stated, there has been much international cooperation on the establishment of oceanographic and radiological models to
define the quantity of radioactive wastes, both high and low, that
would not be suitable for ocean dumping. Much progress has been
made with due caution.

Though theoretical and hypothetical studies have been and are carried
out for radioactive waste ocean dumping, little field works have
commenced to verify these models by elucidating the physical, geochemical, and biological processes controlling the behavior of radioactive waste materials in the deep ocean. Mr. Dyer summarizes the
U.S. effort to survey the waste dumpsites, both in the Pacific and
Atlantic Oceans.

Earlier works reported that within experimental error there was no
radioactivity that exceeded background levels in the water, sediments,
or biota. Recent works show contrary observations. For instance,
Dayal et al.[1] reports that both ^{137}Cs and ^{134}Cs were present in significant concentrations in sediment samples close to a waste canister
in the abandoned 2,800-m Atlantic nuclear waste disposal site located
at 38°30'N and 72°06'W. They report that the observed ^{137}Cs concentrations are orders of magnitude higher than fallout background concentration levels in the area. They further thought that bioturbation
being the dominant mechanism by which cesium is redistributed, their
mathematical model calculation yielded that total ^{137}Cs activity
released into the overlying water column as a result of bioturbation
is a small fraction, 0.3 percent, of the ^{137}Cs released from the
canister.

I concur with Mr. Templeton's summary stating there are clear needs
for additional scientific investigations into processes controlling
the behavior of radioactive materials in the deep oceans and for
radiological surveillance programs designed to provide baseline
data to assist in verifying various models already advanced and the
deep-sea ecological model yet to be developed. Especially, I wish to

emphasize the development of reliable deep-sea ecological models,
for one of our main concerns is the impact of radioactive waste
ocean dumping upon the anthroposphere via the oceanic ecosystem. The
first recipients of the waste impact can be the deep-sea organisms
whose ecology we know little.

DEEP-SEA ECOLOGICAL STUDY

Food for deep-sea bottom communities ultimately comes from the above
at the sunlit region.[2] Recent data show that deep-sea community is
not too different from its counterpart in shallow waters. Faunal composition includes polychaete warms, bivalve molluscs, peracarid crustaceans, foraminifora, nematodes, harpacticoid copepods. In addition,
highly mobile fishes and scavenging amphipod crustaceans are often
witnessed via baited cameras and submersibles, though we still have
difficulty in quantifying them. An emphasis I want to make here is
that deep-sea communities contain a very large variety of species
(Fig. 4.4). Any modelling attempt of such a multi-species ecosystem
at the bottom of deep ocean is a formidable scientific undertaking;
it obviously will necessitate a long-term scientific inquiry.

Because of very stable environmental condition of the deep sea,
organisms there must have enjoyed easy life cycle once they adapted
themselves to their environ. Consequently, they probably can adjust
to only a small environmental change. Thus, any anthropospheric
disturbance, such as radioactive waste dumping and manganese nodule
processing plant reject dumping, can be quite traumatic upon them,
as if fragile greenhouse flowers suddenly transplanted in a garbage
dumpsite. Their recovery from the anthropogenic trauma can be very
slow.

Another consideration we ought to make is that any organic substances,
such as wooden crates in the waste canister, can be potential food
sources. Concurrently, these canisters may provide shelters for
preys against their predators.

Though we are land-dwellers, we have been concerned very much about
the potential transfer of harmful substances through biological food
chains which eventually affect our welfare. Biological shunt,
shortcut, can occur, for instance via vertical migration, placing
seemingly isolated deep-sea community as a part of the dynamic
oceanic ecosystem. Adequate scientific basis must be established to
know of such processes.

Hessler and Jumars reported that nothing is known about the ways in
which deep-sea organisms will respond to added radiation doses.[3]
How could we measure it? In-situ measurement can be preferred, but
we still lack its proper instrumentation. Perhaps our early radiation
dosimetry study will be carried out via captured live deep-sea organisms in laboratory.

Any field testing of the deep-sea ecosystem impact modelling is a
formidable task. The deep-sea community is complex and its reliable
impact prediction, both short- and long-term, can be realized after
much basic scientific researches are carried out. For instance, the
present available data on the deep-sea community, which began during
the Challenger expedition of 1872-1876, are still inadequate to

FIG. 4.4

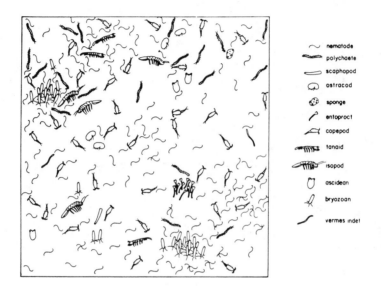

The fauna found in 0.25 square meters of bottom at 5597 meters in the central North Pacific (Station H-153: 28°25.91'N, 155°30.05'W). All of the animals are much smaller than depicted here, such that if they were in true proportion to the square, one would see nothing.[4]

ascertain the dynamics of the deep-sea ecosystem. Yet, the fact is
that we are using the deep ocean as a waste depository. I personally
consider that the deep-sea biological study is the most critical path
on abyssal waste management undertaking of our society.

To begin a systematic quest for the deep-sea ecological study in
reference to waste dumping, I would like to propose that the United
States seriously consider designing a 6,000-meter capacity deep-sea
submersible vehicle with adequate research and experimental capabilities. At present, the United States' Deep Sea Research Submersible
Alvin has a diving capacity of 4,000 meters; it is not useful at the
ordinary deep-sea floor of 5,000 meters. The lessons we have learned
from Alvin would be incorporated in the new vehicle.

Dr. Kilho Park

REFERENCES

1. Dayal, R., et al. Radionuclide Redistribution Mechanisms at the
2800-m Atlantic Nuclear Waste Disposal Site. Deep-Sea Research,
Vol. 26, No. 12A, 1329-1345 (1979).

2. Hessler, Robert R., and Peter A. Jumars. Abyssal Communities
and Radioactive Waste Disposal. Oceanus, Vol. 20, No. 1, 41-46
(1977).

A REVIEW OF FIELD STUDIES AT UNITED STATES DUMP SITES

Robert S. Dyer
Office of Radiation Programs
U.S. Environmental Protection Agency

INTRODUCTION

In this panel discussion on low-level nuclear waste disposal in the
oceans, I will begin by focusing on two points arising from Mr. Templeton's talk. The first point concerns the basic format and utili-
zation of the IAEA Information Cicular 205, Addendum 1, Revision 1
which provides a Definition of high-level radioactive waste or other
high-level radioactive matter unsuitable for dumping at sea (in
response to Annex I, Paragraph 6. of the London Convention), and
Recommendations for controlling the sea disposal of all other radioactive materials at sea (per Annex II, Paragraph D. of the London
Convention). This document contains three distinct sections:

1. The Definition of high-level radioactive waste unsuitable for sea
disposal; this Definition must be observed by all countries (Contracting Parties) who have ratified or acceded to the Convention,

2. Recommendations for issuing permits for sea disposal; these
Recommendations, while not mandatory, should be taken fully into
account, and

3. The Annex to the IAEA Definition and Recommendations which provides background material elaborating upon the considerations used
to develop the Definition and Recommendations. With respect to the
Definition it should be noted that the derived specific activities
of alpha and beta-gamma-emitters, and tritium referenced by Mr.
Templeton are flexible because they are all based on "an assumed upper
limit to the mass dumping rate of 100,000 tonnes per year at a single
dumping site." If the assumed mass dumping rate were to increase or
decrease, the derived specific activities would vary inversely. If
the dumping area were to be increased from a single dumping site to
a finite ocean volume of $10^{17}m^3$, the derived specific activities
would, with the exception of alpha-particle emitters, also increase,
as explained in the IAEA Recommendations.

The second point concerns the form of the Definition. The specific
activity approach is used to define those materials which cannot be
dumped because the release-rate-limit or dumping-rate-limit approach
for individual radioactive materials presents a significant administrative problem. To apply this rate-limit concept would require a
detailed knowledge of the isotopic composition of the radioactive
waste. This in turn would require lengthy and extensive analysis
which could not be justified for material classified as waste.
Normal waste handling procedures segregate radioactive waste into
alpha-emitters and beta-gamma-emitters hence the present derived
specific activity approach requires minimum additional time or expense
to manage. It should be noted that semantically there is no difference at this time between release rate limits and dumping rate limits.

We must assume that whatever is dumped into the ocean is released since no demonstrated containment of packaged nuclear waste has been shown for wastes dumped at the major international nuclear waste dumping area in the Northeast Atlantic.

With that brief digression I shall now turn to the subject that I wish to discuss -- What sort of actual scientific work has been or is being conducted to look at the fate and potential effects of nuclear waste disposed in the deep-sea? The technical guidelines and recommendations that we have been discussing thus far have been developed primarily by advisory panels sitting in large rooms such as this. But what about actual oceanographic work to evaluate this nuclear waste disposal option? What has or could be done? While relatively little ocean dumpsite evaluation work has been done to date for low-level radioactive waste disposal, I would like to discuss the kinds of oceanographic work that have been done in the past, the utility of this kind of work, and some of the preliminary findings that have been made to date.

PAST OCEANOGRAPHIC WORK

There was some survey work done in deep-sea nuclear waste dumpsites used by the United States between 1946-70, but no detailed survey work has been performed in the international dumpsite located in the Northeast Atlantic and in use since 1967 by other countries.

Three studies of the U.S. sea disposal sites were conducted prior to 1970: two on the West Coast at depths of approximately 6,000 feet; and one study on the East Coast over a depth range of 9,000 - 13,000 feet. As I specifically documented in the background paper (Chapter 3), none of those studies detected radioactivity exceeding background levels, consequently no significant dose or health effect to man was implied from the U.S. sea disposal activities at those sites. However, the basic problem at the time of those studies (1957-62) was the inadequacy or lack of sensitivity of the radioactivity measurement systems , as clearly stated in the report of one of those earlier studies. The measurement systems were not really adequate to detect the concentrations of radioactivity that might be expected to be released from packaged low-level radioactive wastes. Another problem was the limited engineering systems capability to conduct on-site evaluations of the fate of the radioactive waste packages and their contents. It is highly likely that much of the survey activity was not actually conducted in the immediate vicinity of the radioactive waste packages hence conclusions regarding the performance of packaging systems and the movement of any released radioactivity were difficult if not impossible to obtain. Today with systems such as Loran C navigation, manned and unmanned submersibles, deep-sea collection and recovery devices, and towed side-looking sonar systems, the capabilities exist for site-specific survey and location of nuclear wastes and, as I shall discuss later, demonstrated recovery of selected nuclear waste packages from depths as great as 13,000 feet.

One of our primary concerns today in evaluating the sea disposal option is to identify and determine the importance of those specific pathways through which radioactive materials might actually move vertically and horizontally from a dumpsite to become a concern to

to man; pathways such as currents and water-mass movements, and biological food chains. The Office of Radiation Programs of the EPA is
concerned with learning as much as possible about past U.S. sea disposal operations in order to determine whether the assumptions regarding the fate and behavior of the radioactive wastes and their packaging systems that have been made during the past 25 years are indeed
correct. This will substantially assist us in developing regulations
and criteria in such areas as site selection, monitoring, packaging,
and site designation to provide environmentally-responsive controls
on any potential nuclear waste disposal into the oceans. I think
the need to verify past assumptions is quite clear. For example,
when we queried participants in the earlier surveys and dumping operations as to the fate of the dumped radioactive waste packages, the
replies ranged from an assumption of no damage to the drums whatso-
ever to an assumption of complete fragmenting of the drums during
or after descent to the sea floor. There was a ready admission that
one really did not know if, and to what extent, the radioactive
waste materials were released. One could only assume that some of
the radioactivity escaped to the surrounding environment.

Therefore, we embarked on a series of site-specific surveys at the
four major U.S. disposal sites to validate, correct, and amplify
past assumptions. We approached these surveys intent on three basic
objectives:

1. to demonstrate that the technology exists to locate radioactive
waste drums in a designated ocean dumpsite, thus insuring that these
sites can be periodically surveyed or monitored;

2. to determine the condition of these drums and their immediate
deep-sea environs; and

3. to determine whether measurable amounts of radioactivity were released.

The question of "monitorability" of any ocean dumpsite is of prime
importance since monitoring is requisite to the continued validation
and refinement of assumptions concerning the fate and behavior of
radioactive materials dumped into the deeper sections of the world
oceans.

What I would like to do now is apply the old adage that "a picture
is worth a thousand words" and describe with a few slides what we at
EPA have learned and achieved relative to the three objectives I just
stated. These slides won't give quantifiable estimates of any
hazards from these dumped radioactive waste drums but they will provide information as to how the radioactive materials may be released,
where they might go, and some of the biological organisms living in
the vicinity of the drums which could ingest released radioactive
materials. A very important concept in exposure pathway assessment
that is receiving increased attention is the biological short-circuit
possibility. This idea has been addressed at advisory group meetings
of the International Atomic Energy Agency. And one of our chief
concerns at this time is investigating the possibility and probability
of such short-circuit pathways. These pathways are not readily identifiable in terms of normal biomass transfers, and arise when unpredictable or aperiodic events occur which could remove released
radioactive material from a dumpsite directly to the surface, shore,

or exploited fisheries.

SUBMERSIBLE SITE STUDIES

The EPA Office of Radiation Programs has availed itself of the dis-
used U.S. nuclear waste dumpsites as valuable study areas for radioactive material behavior as a result of U.S. use of the sea disposal
option for over twenty-five years. We felt that submersible technology could provide the needed site-specific survey capability at
U.S. nuclear waste dumpsites, and that such a capability must be
clearly demonstrated before any future sea disposal can occur. This
led to the first of five EPA site-specific surveys at the four major
formerly-used U.S. ocean disposal sites for low-level nuclear waste.
Each survey used either a manned or unmanned submersible and I shall
briefly describe the submersibles and selected photos taken in the
dumpsite surveillance areas.

In 1974 we used the U.S. Navy's unmanned Cable-controlled Underwater
Recovery Vehicle (CURV III) to provide the first successful search
and location of nuclear waste packages dumped under AEC license between 1951 and 1953. Fig. 4.5 shows the CURV III submersible at the
dumpsite with its sampling arm and sediment coring tubes extending
out to the right of the vehicle, and the electrical control cable
trailing back at the left. The dumpsite is located near the Farallon
Islands about 40 miles west-southwest of San Francisco, California
at a depth of approximately 900m (3,000 feet). In 1975 we conducted
a similar study, using the CURV III, of the companion Farallon
Islands dumpsite located about 50 miles west-southwest of San Francisco, California at a depth of approximately 1,700m (5,600 feet).
These two dumpsites are of particular interest because more than
50 percent of all waste packages and 25 percent of the total curies
of radioactivity dumped by the U.S. was put into these two sites
(see Table I of Chapter 3, "Sea Disposal of Nuclear Waste: A
brief History"). This region then should present us with a high
probability of locating radioactive waste packages if the submersible
system functions adequately.

The assumption was correct and the submersible system worked. Fig.
4.6 shows one of the first radioactive waste drums seen at the 900m
Farallon Islands dumpsite. Lack of adequate labelling was a continuing problem with most of the drums seen in both the 900m and 1,700m
Pacific dumpsites. In this photo the presence of the wire lifting
eye seen protruding at the upper middle of the right side of the drum
(concrete end) indicates that this is a radioactive waste drum dumped
between 1951-1953. The presence of the hydrostatic implosion point
in the upper center of the drum further supports the historical
records describing the packaging characteristics of the drums dumped
in this site. The packages generally consisted of a concrete cap at
each end of the drum with the waste and unwanted air voids sand-
wiched between the caps. In the depression at the top center of the
drum can be seen a deep-sea sole, Embassichthys. This shows quite
simply how a fish might take up radioactivity directly from the waste
drums and, if commercially caught, how it could transfer this radioactive material directly to man. This photo also illustrates to some
extent the artificial reef concept -- both fish and invertebrates are
attracted to large objects, such as waste drums, placed on the ocean
bottom in areas of relatively featureless expanses of soft mud.

Eventually an extensive benthic community builds up around these objects placed there by man. Fig. 4.7 shows another 55-gallon radioactive waste drum at the 900m Farallon Islands dumpsite. The pink
fish resting at the metal end of the drum is a thornyhead,
Sebastolobus. This 900m dumpsite would be precluded from use today
because it would not conform to the IAEA recommended minimum disposal
depth of 4,000m and it also contains the commercially exploited
sablefish, Anoplopoma.

Fig. 4.8 shows one of the 55-gallon radioactive waste drums observed
at the Farallon Islands 1,700m disposal site during our 1975 survey.
It shows the lifting eye and concrete-cap packaging typical of this
dumpsite area and again has no identifying labels. It is interesting
to note that there is relatively little biofouling on this drum with
the exception of a few anemones and a crinoid (feather star) at the
upper right surface of the drum. There is also very little corrosion.
Both of these facts suggest a short immersion time or the presence
of a very effective anti-fouling, corrosion-resistant finish. Fig.
4.9 shows one of the radioactive waste drums at the Farallon Islands
1700m dumpsite. This drum has suffered considerable hydrostatic
implosion because of the air voids in the waste between the two concrete caps. It should be assumed that some of the radioactivity was
released through the resulting space between the concrete cap and the
drum wall. There is considerable biofouling on this drum with the
vestige of an undecipherable yellow identification number at the
lower left of the drum. Fig. 4.10 shows another drum at the Pacific
Farallon Islands 1,700m site. Here we have an example of color
coding. Drums painted red usually had somewhat higher radioactivity
contents than the other drums in a shipment. It would be useful if
we could go back and check this drum against some existing inventory
list prepared at the packaging facility. This would provide valuable
information on such parameters as immersion time, initial radioactivity inventory, and waste form, and it would allow more accurate
estimates of the corrosion rate, matrix degradation rate, and the
percent of radioactivity released over a specified time period. But
the record keeping was not done uniformly and most of the needed
records are not available today. In both the 900m and 1,700m dumpsites, plutonium concentrations above the expected weapons-testing
fallout contributions were detected. However, these concentrations,
while detectable, were not of sufficient magnitude to suggest any
health impacts to either man or the marine environment at this time.

In 1977 we planned a more extensive survey of the Farallon Islands
900m and 1,700m sites to more thoroughly describe their physical and
biological characteristics. Included in these plans was the recovery
of a radioactive waste drum to analyze the metal corrosion and concrete matrix degradation rates after more than twenty years of
immersion. This required a manned submersible. We evaluated the
Deep Quest research submersible, shown here in Fig. 4.11, which is
owned and operated by the Lockheed Missiles and Space Company. In
trial dives to 2,000 feet it demonstrated a capability to perform
the recovery but subsequently was not available at the required time
period scheduled for the survey and recovery operation. We elected
to use the manned submersible PISCES VI owned and operated at the
time by International Hydrodynamics Company of Vancouver, British
Columbia. The logistics and research support vessel R/V Pandora II
was provided through a joint agreement between the U.S. Environmental
Protection Agency and the Canadian Department of the Environment.

Fig. 4.12 gives a full side view of the PISCES VI taken while the
submersible was still about six feet off the bottom. In the foreground is part of the sediment grab sampling system. In the background is an intact 55-gallon drum sitting upright with an orange
fish (Sebastolobus, or thornyhead) resting on top of the drum. A
large hexactinellid or glass sponge can be seen growing on the right
side of the drum with a portion of another specimen visible behind
the fish. Here again the fish seems to favor the drums over the
surrounding mud bottom.

The presence of this sponge raises an interesting question. Can
biota such as this sponge which secretes chemicals to enhance its
attachment ability, significantly effect the deterioration rate of
the metal drums through other actions than the mere change of oxygen
concentrations by encrustation? Fig. 4.14 shows another top view of
a radioactive waste dumpsite. The sediment grab sampling device is
in the foreground. The drums shows the typical concrete cap and
metal lifting eye. Two specimens of a hexactinellid (glass) sponge
are seen growing on the concrete cap. The distal one-third of the
drum shows considerable hydrostatic implosion. The macrofaunal
fouling over the metal surface of the drum is minimal.

In 1976 and 1978 we conducted surveys at the other two major U.S.
deep-sea nuclear waste dumpsites. These are both located in the
Atlantic -- one at a depth of 2,800m (approximately 9,200 feet) 120
miles east of the Maryland-Delaware coast, the other at a depth of
3,800m (approximately 12,500 feet) 200 miles east of the Maryland-
Delaware coast. More than 30 percent of all radioactive waste
packages and 75 percent of the curie totals of radioactivity dumped
by the United States were put into these two Atlantic dumpsites.

We initiated our submersible surveys of these two primary U.S. Atlantic dumpsites in 1975 with a series of three dives at the 2,800
site using the manned submersible ALVIN. Fig. 4.15 will give you
some idea of its size. The front of the submersible is at the left.
The manipulating arm of the vehicle is not attached in this photo.
This is a three-man submersible with a depth capability of slightly
more than 13,000 feet (4,000m) which, coincidentally, is the minimum
nuclear waste disposal depth recommended by the IAEA pursuant to the
London Dumping Convention. Fig. 4.16 is one of my favorite photos.
It is one of the first photographs of a deep-sea radioactive waste
package seen from any Atlantic dumpsite. Here we can see that the
drum is not completely disintegrated; that it has not been completely
destroyed by the great external pressures, although there is clear
evidence of considerable surface corrosion. But we also have to
recognize that the metal drum was essentially just an outer rigid
form whose primary function was to facilitate the handling and transport of the package. The drums were filled with concrete in most
of the east coast disposals. The concrete provided the principal
barrier to the migration of the radioactive materials, and the homo-
geniety of this matrix was therefore desirable to prevent implosion
and waste migration. The concrete in the drum is visible at the
right with a pink anemone attached to it. A small sea urchin,
Echinus affinis, is seen on the top of the drum with two specimens
of the ever-present rat-tail fish swimming over and behind the drum.
There are no signs of implosion of this drum thereby suggesting that
the concrete packaging was quite homogeneous. The sediment surface
of the area around the drum is relatively barren, with the drum

itself providing the major topographic relief. The artificial reef
effect is again in evidence, showing that the few organisms present
in the area were attracted to the drum.

It is interesting to talk to people about their expectations of what
the bottom would look like at this depth. Because of many general
articles on the deep-sea, they expect to see giant fish with huge
teeth, which of course are primarily close-up views of small mid-
water fish. The rat-tails (Nematonurus armatus) seen here are
ubiquitous below about 2,500 to 3,000 meters and we saw them in considerable numbers. And while they may not look too palatable they
are actually quite edible. We recovered a drum very similar to the
one seen in this photograph and it was in very good condition considering that it was immersed for over a decade at pressures in
excess of 4,000 pounds per square inch (psi). Fig. 4.17 shows what
this drum looked like immediately after it was brought to the surface
in the recovery harness. The drum was intact and in surprisingly
good condition. However, I should add that we introduced a slight
bias into the drum selected for recovery. We would not have attempt-
ed to recover a drum that was clearly disintegrating. But, in general, most of the drums we observed in the Atlantic 2,800m dumpsite
were in good condition. One of the distinct advantages of good
labeling, as seen in Fig. 4.18, is to allow one to estimate many
years later the rates of change to a nuclear waste package to include, for example, metal corrosion rate, concrete degradation rate,
biofouling rate, or radioactivity release rates and retention capabilities of any recovered packages. In this photograph we can see
that the information was cast directly into the concrete cap while
the concrete was still wet. Although not quite visible in this photograph, the top line of information tells when this drum of waste
was prepared -- 1961. Subsequent lines tell the dose rate in
millirads/hour at the surface of the drum (40 mr/hr) and at one meter
from the drum surface (3 mr/hr), the most hazardous isotope present
(cobalt-60), the weight of the package (1682 pounds), the volume of
the waste-matrix mixture (9.0 cubic feet), and the package number
(#28). Here was an example of a simple but effective method of identification of each package -- where the information was simply
carved into the still-wet concrete. We are then able to return to
the dumpsite many years later and make some determination of the performance of the nuclear waste packages by answering such questions
as: What release of the nuclear material has occurred? What is the
immersion time of the metal-concrete-waste package? What is the
corrosion rate of the metal? What is the deterioration rate of the
concrete? and, What improvements to this packaging system are required for any potential future disposal activities? Both drum indenti-
fication and record-keeping are going to become increasingly important as more countries begin to dump more material at internationally-
approved dumpsites. The present EPA surveys have illustrated a need
for both uniform and durable package identification, as well as cen-
tralized record keeping for all countries using the sea disposal
option for nuclear wastes.

In 1977, about half way through our site survey program, the IAEA
recommended increasing the international minimum acceptable disposal
depth from 2,000m to 4,000m. Since the only one of the four primary
U.S. nuclear waste disposal sites still to be surveyed was located
at a depth of 3,900m, the capability to survey this deep-sea site
took on increasing importance to generically demonstrate monitoring

and survey capabilities at depths approximating 4,000m. As a signatory to the London Dumping Convention, the U.S. would be expected to
accept technically sound recommendations such as this 4,000m minimum
disposal depth. In 1978 the EPA Office of Radiation Programs carried
out a survey at this Atlantic 3,800m site which is located in the
outer portion of the Hudson River submarine canyon near the axis of
the channel. Again the manned submersible ALVIN was used. The
canyon floor in this region is about one kilometer wide with walls
about 200m high and varying from gently sloping to almost vertical.

Fig. 4.19 is generally characteristic of the area. Here we can see
rounded cobbles probably carried to deep water by ice rafting. Also
present are some brownish white marl chunks probably derived from
slumping off the canyon wall. No fish or invertebrates are seen here.
And the sediment is smoothed by a fairly strong southwest flowing
current. All of the above features of this dumpsite area suggest a
very geologically unstable area. We also found that submersible
operations in the area were quite difficult. There was considerable
turbidity which severely reduced the visibility. The strong currents and steep slope in the canyon axis also made maneuverability
quite difficult. This dumpsite would not provide very good information on radionuclide distributions in sediments around the drums
since most of the sediment surface is scoured and the material is
carried much further downslope. In terms of the ease of monitorability of this dumpsite it would not be a good site for future use.
We recovered a 55-gallon radioactive waste package from this site
as in the 1976 survey at the Atlantic 2,800m site. This package is
still undergoing analysis.

In his opening remarks Mr. Templeton concluded with a statement on
the need for model validation. I think that the validation of
oceanographic models and hypotheses is a very critical stage in
considering the sea disposal alternative. We now have a good starting technical framework or set of recommendations from the IAEA for
guidance to the international community in evaluating the sea disposal alternative. However, no matter how good the information is
today, I think that the oceanographic community would be the first
to recognize that it will be even better two years from today, and
two years hence from that day. Therefore, continual review and
updating of earlier recommendations is very important. The most
important means to accomplish this is to make field measurements
(monitoring and surveys) to validate, refute, or refine previous
assumptions.

CURRENT CONCERNS

Among the factors that we are most concerned about today are biological food chains and specifically the possibility of short-circuit
mechanisms. Another concern and goal is to better define the
currents and physical transport mechanisms, and the potential for
transport to man of released radioactivity from a dumpsite at a depth
in excess of 4,000 meters. Still another concern and goal is to
develop a methodology for measuring effects on the marine biota from
radioactivity released at a dumpsite; and to identify and quantify
marine pathways so that realistic quantitative assessments of radiation dose to man can be made rather than the very rough estimates
of radiation dose presently being made. Any attempt to quantify

actual radioactivity transport by currents or marine organisms from
a dumpsite through the water column to specific marine or human
critical population groups would be extremely difficult, if not impossible, at this time. A dedicated effort will have to be made to
identify intermediate transfer levels, short-circuit mechanisms,
and other potential means by which radioactivity could reach critical
human or marine populations. This must be done well in advance of
any serious consideration of sea disposal as a viable option for
U.S. low-level nuclear waste disposal. Obtaining the needed information will be a long and laborious process. The final result must
be an excellent predictive capability and validation system for
measuring the effects of nuclear waste disposal in the ocean. No
matter how good an oceanographic model is today, there is room for
improvement. And the predictive capability for determing the effects
is directly related to the technical adequacy for any model which
is, in turn, related to the adequacy of the oceanographic information
base.

An important closing thought which I have touched upon in the
previous photographs is that if, in the future, we commence placing
large numbers of radioactive waste packages on the seabed in sparse-
ly-inhabited, deep-sea regions, we could succeed in attracting the
very same organisms we sought to avoid. This could occur through
the same artificial-reef attracting behavior exhibited by fish and
other marine organisms when old cars or tires are placed in shallow
water coastal areas.

From our initial surveys at the U.S. ocean dumpsites we can conclude
that the technology exists or can be improved to properly evaluate
the on-site results of deep-sea nuclear waste disposal operations.
The formerly-used U.S. ocean dumpsites for nuclear waste can provide
key study areas for determining both packaging performance and
radionuclide transport processes. And further effort should be made
to determine the specific transport mechanisms operative at these
sites with their varied depths, currents, topography, geology and
biota.

Fig. 4.5

Fig. 4.6

Fig. 4.7

Fig. 4.8

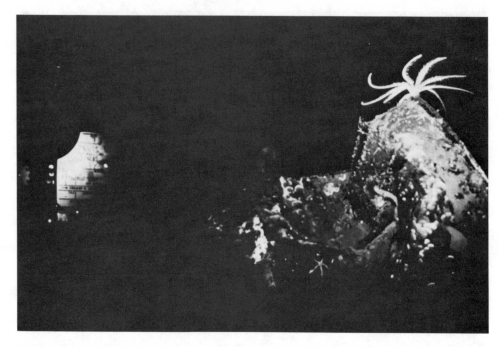

Fig. 4.9

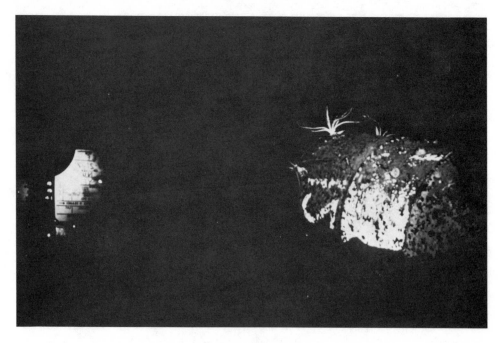

Fig. 4.10

Fig. 4.11

Fig. 4.12

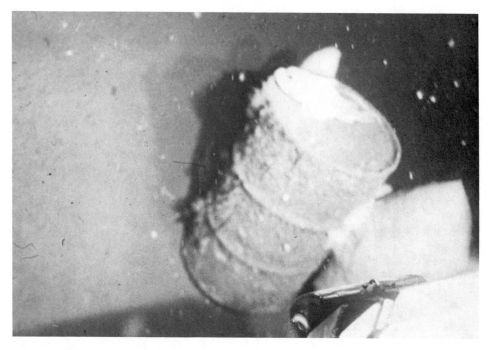

Fig. 4.13

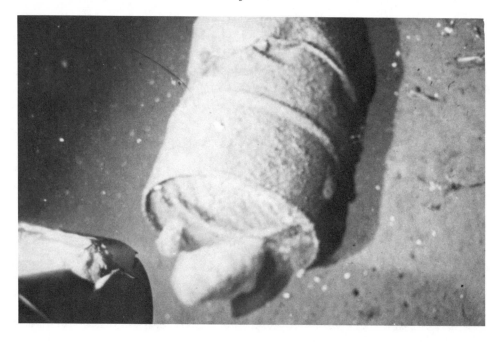

Fig. 4.14

Fig. 4.15

Fig. 4.16

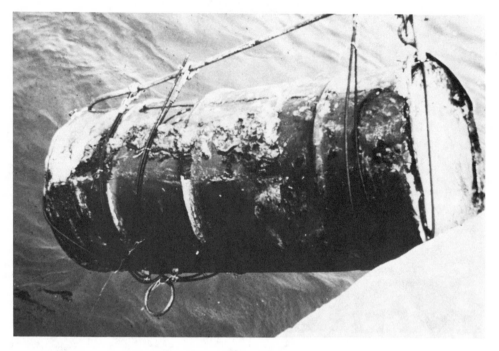

Fig. 4.17

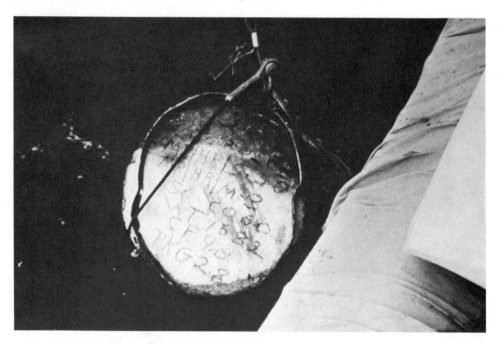

Fig. 4.18

Fig. 4.19

QUESTION AND ANSWER PERIOD

MR. ROOSEVELT: Thank you, gentlemen. We will spend the next 10
minutes on questions and answers involving the audience.

QUESTION: Just before I left San Francisco, I had two different
phone calls from people who were concerned about the possible effects
that the recent earthquake would have on the Farallon Islands nuclear
waste dumpsites because of the San Andreas Fault running near those
sites. I think that this is a reflection of public concern in
general on what might be the effects of the radioactive wastes in
these dumpsites and I wonder whether you could talk a little about
this. I know that you did some survey work in the Farallon Islands
dumpsites during the summer of 1977 and that you also analyzed some
fish and other biological organisms for radioactivity, the results
of which, I understand, you have just received. Could you comment
both on the implications of the proximity of the Farallon Island
dumpsites to the San Andreas fault line as well as the results of
your analyses of the fish and invertebrate samples?

MR. ROOSEVELT: I assume that this question is directed to Mr. Dyer.

MR. DYER: In my earlier discussion I mentioned the IAEA Recommendations for controlling sea disposal of nuclear waste. In these Recommendations one of the requirements for consideration in selecting a
dumping site is that it should not be situated in areas of known
natural phenomena, such as volcanoes. Clearly this would also
include areas of geologic instability such as fault lines. The
Farallon Islands dumpsites do not meet that IAEA requirement and
would not be suitable for future use just based on that criteria
alone. If I might rephrase your question, what, if anything, would
happen to the drums in the event of a severe earthquake in the dumpsite region? First of all I would estimate that up to 25 percent of
the drums of nuclear waste have already released their contents while
the other 75 percent have slowly been releasing their contents to
varying degrees through leaking or leaching processes. In the early
dumping operations the release of the contents from the drums was
expected. This provided for the dilution and dispersion of the
material. However, all of the radioactive material was and is still
undergoing radioactive decay thus reducing the hazard of the material
to a greater or lesser extent, depending upon the half-life of the
isotope.

Therefore, in the event of severe tectonic activity in the dumpsite
area, there would be no catastrophic release of the entire inventory
of the drums. The expected scenario would be that the entire package
itself would be moved around and some would be ruptured but this
would not result in any appreciable additional release. However, in
terms of actual inventories of already-released material presently
absorbed to the sediments, we could expect a significant movement
of this material. But this would not appreciably change the uptake
of radioactive material through the food chain since the fish and
other organisms would neither eat much more sediment as a result of
tectonic activity nor would their bioaccumulation factor be significantly altered. The major alteration which could occur might be an
increase in the amount of radioactivity-contaminated sediment trans-

ported shoreward. But again as I pointed out earlier, the concentrations of the radioactive materials are at such low levels that in 1958 and 1962 they were indistinguishable from natural background concentrations within the limits of available instrumentation. At that time, the instrumentation was only measuring gross alpha or gross beta activity while today we can detect specific radionuclides such as plutonium or strontium if they are even minimally present in the sites. Could you please repeat your other question?

QUESTION: I was under the impression that your recent work gave you at least detectable levels of radioactivity in some organisms found in the water columns.

MR. DYER: I believe your question refers to some fish that we volunteered to analyze for the Point Reyes Bird Observatory which is located on one of the Farallon Islands. We have recently received a draft report of the radioanalytical results. We did find some radioactivity in selected fish but only bottom-dwelling or bottom-feeding specimens, not those that reside in the water column. We found that the two specimens of dover sole that we analyzed showed small amounts of plutonium-239 in their gastrointestinal tracts, but that this may be associated with the high volume of ingested sediment also found in their GI tracts. Two ling cod specimens showed no observable radioactivity except Cs-137, which was found in various body parts at the upper levels of expected concentrations attributable to weapons-testing fallout alone. A squid collected from the water column directly above the dumpsite, during the 1977 Survey, showed no radioactivity other than naturally occurring potassium-40 and radium-226.

An observation I have noted over the past few years is that the public concern over plutonium in marine organisms is greater than for cesium. Yet plutonium uptake is discriminated against in marine organisms such that it moves primarily through the GI tract while cesium is bioaccumulated in the edible portions, e.g. muscle, of marine organisms and could therefore pose a greater risk in food-chain transfer mechanisms.

Further measurements of radioisotope concentrations in various representative trophic level organisms from these dumpsites could provide some very useful information on:
  1. measured bioaccumulation factors,
  2. whether these values verify the postulated bioaccumulation factors, and
  3. whether some inferences about specific trohpic level transfer mechanisms could be drawn from these measurements.

QUESTION: Can you give us some impression as to the levels of radioactivity in the sediments outside of the dumpsite area? Are there detectable levels that are the same or different from the sediment concentrations in the dumpsites.

MR. DYER: In the two Farallon Islands dumpsites the principal radioactive contaminant was plutonium. Chemically, plutonium binds or absorbs to the sediment in the marine environment and therefore would not be expected to migrate long distances unless the sediment itself was transported. Our preliminary measurements in this area have shown plutonium-239, 240 concentrations immediately around the drum

packages in the 900m dumpsite ranging from 2-25 times higher than
the maximum observed weapons-testing fallout concentration for this
latitude and depth of water. Similar measurements in the 1,700m
dumpsite showed concentrations only up to about four times the maximum observed weapons testing fallout concentration for this latitude
and depth of water.

In both sites these concentrations dropped off sharply to background
levels within a few meters of a drum cluster indicating limited
migration of the released plutonium. Therefore even within wide
and poorly-defined dumpsite areas the plutonium concentrations fell
below statistically meaningful detection limits, indicating only a
localized type of release not uniformly distributed throughout dumpsite areas. If we consider:
  1. the apparent limited transport of released plutonium,
  2. the artificial-reef type of biological attraction for the
      drums, and
  3. the fact that the only plutonium we found in the fish analyzed
      for the Point Reyes Bird Observatory was in the GI tract,
it would appear that any easily detectable plutonium found in dumpsite organisms would probably result primarily from ingestion of a
plutonium particle into and through the GI tract from feeding on the
sediment in the immediate vicinity of a drum which has leaked. More
importantly, unless one happened to catch these fish from the immediate dumpsite vicinity and then proceeded to eat these fish whole
without removing the entrails, it would be unlikely that any measurable plutonium ingestion by humans would occur as a result of eating
these bottom-feeding fish.

Based on the levels of radioactivity we have measured in these two
West coast dumpsites so far, the primary significance of the radioactivity concentrations is not as a health hazard but as a tracer or
tag to follow the mechanisms controlling such movement. Radioactive
tracers have long been used in the laboratory to study isotope
behavior. And here at these dumpsites we also have an opportunity to
study plutonium and cesium isotope behavior in a natural environment
utilizing an existing source.

5. SUB-SEABED DISPOSAL OF HIGH-LEVEL NUCLEAR WASTES

INTRODUCTION

Disposal of high-level nuclear waste in the geological formations of
the deep seabed is an aspect of the "ocean alternative" which has
received a significant amount of attention in the scientific community. Currently, this option is being studied through DOE's Sub-seabed
Disposal Program. In this presentation, Dr. Charles Hollister discusses some of his work and the findings of this effort.

Dr. Hollister reviews general criteria for determining the suitability of nuclear waste repository sites in general terms before turn-
ing to the ocean environment. After dividing the ocean floor into
three geological regions, he discusses characteristics of the continental margins, ocean basin floor and mid-oceanic ridge in the context of nuclear waste disposal concerns. Dr. Hollister lists unsuitable areas and introduces the concept of a "multiple barrier"
approach in the study of this alternative.

The Sub-seabed Disposal Program is placed in the context of DOE's
National Waste Terminal Storage Program by Glen Boyer. DOE is
charged with developing methods for disposing of nuclear wastes in a
safe, environmentally acceptable manner.

To date, DOE has concentrated its attention on land disposal options,
evaluating prospects for using conventional mining methods to dispose of high-level wastes in stable geological formations such as
rock salts and granite, while keeping the seabed alternative open
for study and discussion, Mr. Boyer notes.

Sub-seabed disposal should be seen as a concept, not an alternative,
in the process of managing America's nuclear waste, according to
William Bernard. Additional information is needed on topics which
range from the form of wastes suitable for sub-seabed disposal to
retrievability, thermal transfer, storage, transportation, and costs.
Despite these unanswered questions, the Sub-seabed concept has, in
some ways, progressed farther than other options. Mr. Bernard ends
his presentation with a warning against siezing the concept of sub-
seabed disposal as a quick fix for national nuclear waste policy
questions, noting that in many cases our own experience in managing
toxic wastes has shown "haste does make waste."

Discussion after this panel included a review of the questions facing
oceanographers studying the sub-seabed; a question on the cost of
this methodology; and a delineation of areas under study in the Sub-
seabed Disposal Program. (Editor's note).

Fig. 5.0 Charles Hollister

A REVIEW OF CURRENT SCIENCE AND TECHNOLOGY FOR DISPOSAL
OF HIGH-LEVEL RADIOACTIVE WASTES WITHIN GEOLOGICAL
FORMATIONS OF THE DEEP SEABED

Dr. Charles D. Hollister
Dean of Graduate Studies, Senior Scientist
Woods Hole Oceanographic Institution

I am going to start with a brief period of reminiscing. My main
research interest has to do with the procession of sediments on the
deep bottom of the ocean, not the disposal of nuclear wastes.

My interest in radioactive waste disposal at sea stems from a question which was posed to me eight years ago: "We are running out of
backyards and the ocean's geology may comprise the last of this
planet's formations that we've got and we have a lot of nasty stuff
like radioactive materials to get rid of and what do you think
about it?"

My first reaction was "Not in my ocean. You could put it in someone
else's ocean but not in my ocean."

Then the question was posed: "Well, tell me why not?"

And then the scientist takes over. You have to think of very good
reasons other than "No, not in my ocean." You really have to say
why not. Or why. That is the job of the scientist. As one begins
to wonder "why not" -- or why -- one is pushed to think. Then you
look at the oceans with insights that we have gained in oceanography
during the past 100 years, and particularly during the last two or
three decades, and particularly in view of our understanding of how
plates move; at what speeds and directions they move.

Basically, the earth's crust moves in predictable directions and
speeds. How can we use this predictability? Especially in our
efforts to dispose of nuclear wastes? So now we're going to walk
through my Marine Geology 101 course. Those of you who have already
taken it can go back to sleep. Those of you who haven't should be
prepared to ask questions because it is through this kind of dialog
that we have developed and honed our thought processes concerning
high-level wastes and the oceans.

I have given this talk nationally and internationally to oceanographers. We have distributed all of our published literature
widely. And we are still looking for the reason why we should not
continue to study this particular option. We are looking for sound
technical reasons. I will not argue with you about the need for
power, or the need for weapons. But I will tell you something about
the oceans and the geological formations below the sea floor.

We are not talking about dumping nuclear waste at sea and we are not
talking about putting it on the sea floor. We are talking about
using the geological formations in the same way the United States,

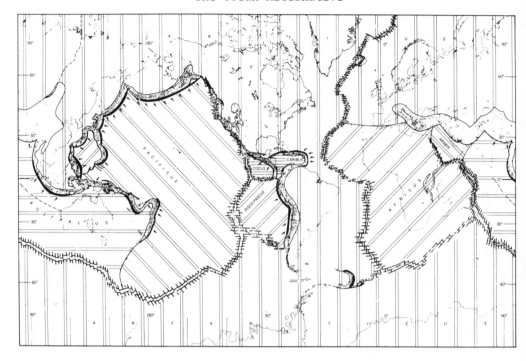

Fig. 5.1

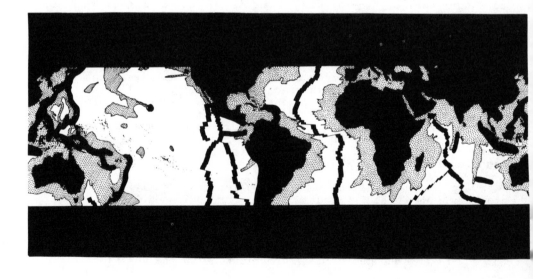

Fig. 5.2

and others, will be using the geological formations on land, as the containment media. We do not believe that man can make a can that will last hundreds of thousands of years. You may argue with me but you have to show me.

INTRODUCTION

Projections of energy demands for the year 2000 show that nuclear power will likely be one of our major energy sources. But the benefits of nuclear power must be balanced against the drawbacks, including its by-product: radioactive waste. While it may become possible to completely destroy or eliminate these wastes, it is most likely that we will have to dispose of some dangerous waste on Earth in such a way as to assure their isolation from man for periods on the order of hundreds of thousands of years.

Undersea regions away from plate boundaries (Fig. 5.1), continental margins (Fig. 5.2), and the relatively productive edges (Fig. 5.3) of major current gyres offer some conceptual promise for waste disposal because of their geologic stability and comparatively low organic productivity. The technical feasibility of permanent disposal beneath the deep sea floor cannot accurately be assessed with present knowledge, and there is a need for thorough study of the types and rates of processes that affect this part of the earth's surface. Basic oceanographic research supported by the U.S. Department of Energy, aimed at understanding these processes is yielding answers that apply to this societal need.

The general requirements for a nuclear waste repository <u>anywhere</u> are:

1. The area must have a history of stability (from 10^{6-7} years ago through the present).
2. It must be isolated from man's unpredictable activities and thus must not interfere with established or likely resources.
3. Disposal is assumed to be permanent, yet emergency retrieval options should be developed.
4. Above all, the site plus the disposal method must act to preclude any significant release of the radionuclides while it is dangerous to the biosphere.

REVIEW OF THE OCEAN ENVIRONMENT

For those not familiar with our present understanding of ocean processes, we will attempt to set the environmental stage for later discussions of radioactive waste repository siting (see also Bishop & Hollister, 1974). Within the ocean areas there are cataclysmic events that could adversely affect any disposal system. They include earthquakes; volcanism; turbidity currents; (See Fig. 5.2, Blackareas), slumping and liquefaction and sediments; erosion by currents; intrusion by molten igneous rock; and intrusion by man. Some are indigenous to the ocean environment; others are characteristic of the earth at large. There are some ocean areas that, on the basis of oceanographic data, appear relatively untouched by any of these events.

The oceans and their floors as they exist today are geologically

young; the oldest oceanic rocks recovered from the deep ocean floor
are less than 200 million years old. Yet the continents contain
rocks of nearly 4 billion years in age, and exposed outcrops of
ancient oceanic sediments now on land, seen by us as old mountain
ranges, are tens to hundreds of millions of years in age.

In the past decade, a revolution has taken place in the concept of
sea-floor evolution. This concept, once called continental drift,
has been reborn as "plate tectonics." It states that the globe is
covered by about a dozen solid-rock plates composed of oceanic or
oceanic-plus-continental crust (Fig. 5.1). These plates move in predictable directions and speeds, and collide in seismically active
regions (Fig. 5.2).

Plate boundaries are either areas of crustal destruction, where the
edges of plates are being thrust under (subducted) other plates, or
they are areas of construction where, if the earth's diameter is to
remain constant, new crust must be made at a rate equal to the destruction rate. This growth takes place along the middle, or rift
valley, of the Mid-Oceanic Ridge (MOR). This feature is a globe-
circling volcanically active welt about 40,000 km long where new
molten basalt is constantly being extruded (Heezen & Hollister, 1971).

It is within this framework of crustal unrest that we describe the
principal features of the ocean environment, highlighting the
characteristics that may apply to the problem of long-term or "ultimate" waste disposal. We do not attempt to discuss the complex biological communities in the ocean provinces, but merely call attention
to the overall biological productivity of surficial waters (Fig. 5.3)
as an important indicator of mid-water and bottom biological activity.

The ocean floors are divided into three principal physiographic provinces, each occupying about a third of the world's ocean area:

1. <u>Continental margins</u> (continental shelf, inland seas and bays,
   marginal plateaus, continental slope, continental rises)
2. <u>Ocean basin floor</u> (abyssal plains, abyssal hills, oceanic
   rises, deep-sea trenches)
3. <u>Mid-Oceanic Ridge</u> (ridge flank and crest, rift valley and moun-
   tains, transform faults)

It is helpful to remember that, with some obvious exceptions, geological processes on the continents are for the most part erosional,
whereas in the oceans they are principally depositional.

The <u>continental margin</u> is one of the most dynamic ocean environments,
with wide seasonal water temperature changes, variable chemical biological processes, and complex and unpredictable geological structures. Here lie most of the remaining unexplored pools of hydrocarbons as well as most of the world's great fishing grounds. Sediment accumulation can be rapid, and this, combined with relatively
steep slopes, provides optimum conditions for sediment failure
(slumping, turbidity currents, etc.). These dynamic elements sufficiently demonstrate that the continental margins do not answer our
criteria of isolation and stability.

The <u>ocean basin floor</u>, occupying another third of the ocean area, is

the deepest (5-11 km) of the three provinces and includes the flat
abyssal plains, the gently rolling abyssal hills, and the deepest
parts of the oceans: the deep sea trenches.

The abyssal plains (gradients of less than 1:1000) contain numerous
deposits of coarse continental debris swept here by periodic rapid
(hundreds of cm/sec) underwater avalanches or turbidity currents.

The abyssal hills, which form broad low swells (mid-ocean rises) lie
seaward of the abyssal plains and are composed of old ocean crust
originally formed as extrusions of pillow basalt from the MOR spreading welt. As the crust spreads away from the MOR it cools and sinks,
reaching a depth below the water surface of five kilometers in about
50 million years. These vast abyssal-hill provinces (e.g. most of
the North Pacific) are generally covered with 10's to 100's of
meters of chocolate-brown clay. Most abyssal hills and rises are
seismically passive. Where they occur below the centers of wind-
driven surface current gyres, they are typically quite stable and
relatively unproductive biologically ("marine deserts"). Bottom
currents are generally weak and variable. From the stability view-
point, these enormous ($10^8 km^2$) regions of mid-plate/mid-gyre abyssal
hills are unparalleled.

The dynamic deep-sea trenches form the landward boundaries of abyssal hills near the collision zones of the sub-sea plates, especially
in the Pacific and Indian oceans. Here the ocean crust is presum-
ably being overridden by lighter crustal rock at rates of two to six
centimeters per year, with attendant crustal destruction. Many high-
intensity earthquakes occur in or near these trenches, triggering
massive submarine slides and sometimes extensive volcanism. The
arcuate circum-Pacific trenches are some of the most dynamic and
least predictable regions on the planet.

The mid-oceanic ridge is a sea-floor spreading welt that runs around
the globe -- a "constructive" plate boundary. Along this symmetri-
cally expanding ridge is a hot, seismically-active rift valley where
new crust is continually being extruded. The temperature of the
central portion of this rift valley may approximate that of molten
basalt ($1,200^oC$), yet water circulation dissipates the heat so quickly that no rise in water temperature can be measured at the sea
surface directly above this tremendous ($20,000 km^2$) heat source.
Sediment in this, the shallowest and youngest part of the ocean, is
generally too thin to be detected except as a current-winnowed
veneer of carbonate shell debris seen in bottom photographs (Heezen
& Hollister, 1971).

With this very oversimplified view of oceanic process as background
we attempt below to define a process of repository site identification that must first focus on generic site suitability criteria.

SITE SUITABILITY CRITERIA

Our fundamental working assumption is that the "host medium" or geologic formation remains the single most important barrier to the
release of radioactive material to the biosphere. The degree of retention is governed by two primary characteristics -- high sorption
capability and low permeability.

Charles Hollister

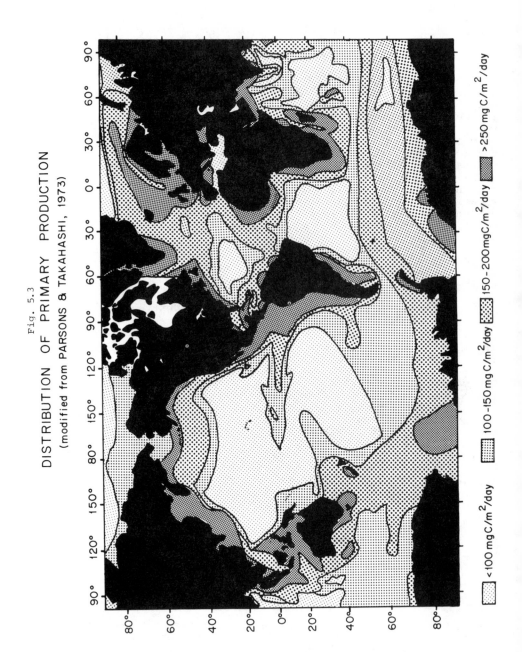

Fig. 5.3
DISTRIBUTION OF PRIMARY PRODUCTION
(modified from PARSONS & TAKAHASHI, 1973)

$< 100 \, mgC/m^2/day$ $100 - 150 \, mgC/m^2/day$ $150 - 200 \, mgC/m^2/day$ $> 250 \, mgC/m^2/day$

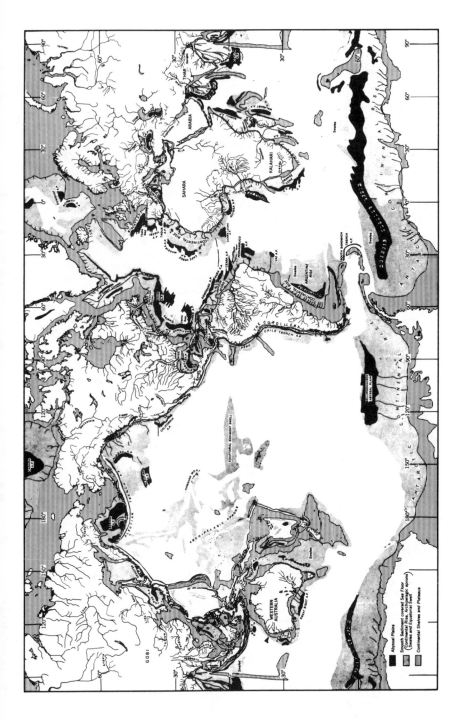

Fig. 5.4

The distribution (or soption) coefficient (Kd) of the host medium is a measure of its ion uptake capacity. In other words, Kd is simply the ratio of the amount of material (radioactive ions in this case) bound to the medium vs. the amount free to travel by diffusion or advection through the pore water. A Kd of one implies that for every ion locked up one is mobile; a Kd of 10^6 implies that the ratio is a million locked up to one mobile. Kd values are primarily related to grain size; i.e. fine-grained sediments have the highest Kd's. A high Kd for a broad range of radioactive ions is one of the key desirable characteristics of a disposal medium.

Another important characteristic is permeability: the measure of the speed with which fluids can migrate through a medium. For waste retention, low rates of flow are desirable. Permeability should not be confused with porosity which is simply the measure of pore space in a medium. The important difference is that a medium such as fine-grained clay may be very porous but have a very low permeability because the pores are not connected. Conversely, a medium such as silt or sand may have a comparatively low porosity but be highly permeable if its fabric includes a large fraction of interconnecting pore spaces.

Another desirable characteristic of the medium is that it behave plastically or visco-elastically to enhance its integrity if disturbed. Such plasticity could also produce self-healing of an emplacement hole.

The role of organic material in processes of remobilization is unknown due to the difficulty of modeling the effect of organics in complexing ions, thus preventing their soption by the sediment. Our initial studies have focused on sediments with the least amount of organic material.

Another even less well understood (or quantifiable) characteristic of the medium is its ability to remain undisturbed under high or variable thermal stress. That is to say, it is necessary to understand the medium's behavior under a variety of thermal fields in order to predict its response as a function of temperature and time. A fundamental question here is whether the existence of a heat source in the submarine repository could cause the sediment to convect and thereby breech the sedimentary barrier. However, canisters could be designed to contain, i.e. by dilution with an inert material, waste until a lower heat content level is reached.

In summary, the most desirable characteristics of the "host medium" are:

1. high Kd and low permeability
2. ability to self heal, i.e. be visco-elastic in response to dynamic stress
3. a low organic matter content (well oxidized)
4. stability under predicted thermal loading.

The submarine geologic formation that appears to best satisfy the above criteria is abyssal "red" clay. Depending on organic interactions and permeability considerations, light brown deep-sea clays 20 to 40 percent $CaCO_3$ may also be suitable. Organic-rich, more permeable biogenic oozes appear less suitable, with turbidite sands and silts least desirable of all. As a first approximation, the white areas of Fig. 5.4 generally satisfy these criteria.

Undesirable examples of host media include:

1. Coarse-grained deposits, such as the proximal portions of abyssal plains (including all fracture zone floors) and ponded turbidite sands of inter-mountain valleys, which are expected to have high and variable permeabilities and are least likely to behave plastically. This criterion effectively eliminates all submarine canyons, continental shelves, shallow portions of the Mid-Oceanic Ridge and fracture zone floors (e.g. less than about 3,000m).

2. Deposits exhibiting high or variable thermal conductivity, or where heat flow gradients are highly variable. These regions lack coherency of thermal properties and probably exhibit variations in geotechnical properties such as bulk permeability. It is too early to review all heat-flow data in order to exclude areas on a global scale; rather it seems prudent to do the exclusion exercise with other more general criteria and then review heat-flow data and conduct special studies in candidate regions.

3. Deposits with high organic carbon content, which are likely to exhibit nonpredictable behavior in the presence of radioactive isotopes. This criterion would probably eliminate the continental rise and slope where organic carbon content in the upper 10's of meters of hemipelagic material ranges between one and 10 percent.

4. Thin or discontinuous sediment blanket. From initial calculations and from experience with the free-falling Giant Piston Corer (Hollister, Silva, and Driscoll, 1973), it appears that a projectile could penetrate 20 to 40 meters into the types of sediment that fit the above selection criteria. Because of the high (and often variable) bulk permeability values that have been determined for at least the upper layers of oceanic crustal basalt, a buried container should be 500 meters or more above basement.

THE GEOLOGIC ENVIRONMENT

The geologic setting or environment within which the medium resides can also be constrained based on certain working assumptions. The most important constraint is environmental predictability over the minimum time span needed for waste isolation (i.e. 10^5 years). The setting should be so well understood (or so simple) that the probability of any natural event altering the repository could be estimated with considerable certainty.

The obvious approach has been, and still is, to seek areas of low geologic activity: tranquil regions with very low levels of seismic and volcanic (tectonic) activity and a long history of continuous deposition rather than erosion. Such regions occur in the centers of large lithospheric plates and beneath the centers of surface circulations gyres that contain the least productive (biologically) and least energetic ocean water.

Thus one site exclusion criteria is proximity to the edges of lithospheric plates. A rule of thumb excludes those areas within 100 miles of any recorded seismic event greater than a given magnitude[2] or within 100 miles of volcanoes known to have been active in the past 10^{6-7} years. All deep-sea trenches, mid-ocean rift valleys, the crests of the Mid-Oceanic Ridge and zones of transform faulting are excluded by this rule.

Another geologic perturbation is recurring glacial ice. Present trends suggest that another ice age will occur within the next 5,000 to 50,000 years. Because this is less than the 10^5 year isolation period, regions likely to be adversely affected by another ice age should be excluded. These areas include all regions known to have experienced ice-age erosion and all deposits containing coarse, ice rafted debris.

Man will seek any natural resource of value, so all regions containing known or suspected concentrations of food, minerals or hydrocarbons should be excluded. From the standpoint of future hydrocarbon recovery, entire continental margins including deep sea fans, cones and aprons should also be excluded. In addition, regions of high biological productivity (such as upwelling areas) which support major fisheries and manganese nodule deposits rich in nickel and copper are excluded also. In the event that these latter surficial deposits could be mined in advance of a disposal operation, relaxing of the criterion might result.

In addition to the above considerations, there are certain technical criteria for low-level waste that are (or soon will be) governed by the IAEA recommendations (IAEA INF CIRC/205/Add. 1/Rev. 1, August, 1978) and should be taken into consideration. Though largely covered above they are listed here for completeness:

1. Sites should like between 45° North and 45° South latitude to avoid sources of bottom water (which are characterized by strong vertical mixing), and the areas of high biological productivity in the polar regions;

2. Depth at the site should be 4000 meters or more where biological, chemical, physical and topographical gradients are generally low, bottom water circulation is slower, and organic carbon in the sediments tends to be low;

3. Sites should be remoted from continental margins to avoid regions of high biological productivity, active resource exploration and exploitation, and geologic unpredictability and instability (continental slope, rise and associated fans and canyons);

4. Sites should be away from areas of potential seabed resources;

5. Sites should be away from transoceanic cables in use;

---

2   This value will be established after completion of our geotechnical/dynamic response program.

6. Sites should be away from areas where geologic hazards such as
    submarine slides, volcanoes and earthquakes decrease a site's
    environmental predictability;

7. The area of a site should be defined by precise coordinates,
    with an area as small as practicable;

8. If possible, sites should be in areas covered by precise naviga-
    tional aids to assist in relocating the site;

9. Sites should be away from features such as submarine canyons
    which may unpredictably affect rates of exchange between deep and
    surface waters near the continental shelf;

10. Sites should be chosen for convenience of disposal operations and
    to avoid, so far as possible, the risk of collision with other
    traffic and undue navigational difficulties; and

11. Bottom current shear stress should not exceed critical erosional
    shear stress to prevent high rates of resuspension and eroding
    of the sediments.

REJECTED OPTIONS

The deep trenches and subduction zones where one plate is being
driven under another have been suggested as a disposal location: why
not put the high-level waste into the deep-sea trenches where they
will supposedly be carried deep under the earth? The deep-sea
trenches are unpredictable; material from trench floors has in fact
been thrust up onto the continent in certain areas. Also, subduction
rates (a few centimeters per year) are too slow to be of any use
during the hazardous period of the waste. Trenches are usually near
continents (and humans) and they generally lie beneath biologically
productive ocean waters. Another, perhaps minor, consideration is
that at present we do not have the technology for penetrating crustal rock or sediments at trench depths.

The crustal rock itself (as opposed to the overlying sediments)
could be a candidate medium. The only direct data we have about the
structure and composition of crustal rock have come from a few holes
drilled with great difficulty through approximately one half km of
basalt. Core samples taken on the Mid-Oceanic Ridge suggest that
this rock is broken up and badly fractured, with perhaps very high
bulk permeability. All of the data so far suggest that shallow
ocean crustal rock is not monolithic. These considerations lead us,
at least for the present, to the thought that emplacement of wastes
in the crustal rock at shallow depths would not be prudent. An
abyssal mid-plate region of old crust should be drilled, however,
before we finally abandon this disposal option, as it may be that
crustal rock is effectively healed and sealed -- leading to a very
low permeability -- by the time it reaches mid-plate depths of 5 km
(a journey requiring at least 50 million years of sea floor spreading).

Placing the high-level waste on top of the sea floor simply by
kicking a canister off the fantail effectively puts the waste directly into the biosphere, unless we can construct a canister that would
survive in the corrosive marine environment for hundreds of thousands

of years. Any leak, either during the disposal operation or after,
would inject <u>some</u> radioactive material into the marine ecosystem.
From samples, photographs, and current meter data, we know that the
energetics of the biological and physical processes at the sediment/
water interface (benthic boundary layer) can be very high and very
unpredictable.

Another suggested disposal option is to dilute the high-level waste
by dispersing it into the ocean waters. Calculations show that the
waters of the ocean are not vast enough to take all of the waste
from military and industrial sources accumulated over the next few
decades without being eventually contaminated beyond safe limits.
In addition, the dispersion and concentration mechanisms -- biological, physical and chemical -- are so poorly known that researchers
are not yet ready to predict possible pathways and rates of transfer
from the surface of the ocean bottom to man's food chain.

A MULTIPLE BARRIER APPROACH

To place reasonable limits on this discussion, we assume that waste
material will be solidified, encapsulated, and emplaced well below
the sea floor. The system to be considered, then, includes the encapsulated waste, the surrounding geologic medium, the overlying
water column and the biological community. We include all natural
rocesses that occur there, plus new processes that may result from
the introduction of radioactive wastes.

We have found it useful to list possible barriers to a release of
radioactivity. We define a release as the set of circumstances
whereby toxic material can reach man or when administrative controls
must be placed upon an area to prevent exposure. The barriers are:

Generic
    Distance from habitation
    Water depth
    Constant conditions (temporally and geographically)
    Geologic stability and predictability (lack of cataclysmic events)
    Sparse biology or low potential for biologic transport
    Large liquid dispersal medium (as last resort)

Stepwise
    Solid, low leach rate
    Corrosion resistant
    Low permeability
    No advection of water through sediments; diffusion only
    High sediment ion-exchange or fractionation
    Special chemistry or biological activity at sediment surface (the
        benthic boundary layer)
    Slow currents (tidal only; very small throughput)

The "generic" category merely describes the setting -- conditions to
which the engineer can design the system. The "stepwise" barriers
are possible effective preventions for dispersal of radioactive
material. For example, several solids with very low solubilities and
leaching rates (approximately 10^{-5} g/cm^2 per day) have been made from
stimulated high-level radioactive wastes, and there are several cask
materials (e.g., nickel, zirconium, titanium) that might have consid-

erable lifetimes in contact with either water or the solid materials
of the rock and sediments.

Several natural accretionary processes occur in or on the sediment;
these could provide added protection by the formation on the canister surface of an additional barrier to movement of the radionuclides.
Water movement upward through the sediments is slow, at rates compa-
rable to those of sediment deposition above it, and the sediments
also retard the movement of some elements by ion-exchange processes.

In case of accidents, which would allow release of radionuclides,
the water column itself would add additional protection through
dilution. Hundreds to tens of thousands of years are needed for
transport of a particle of water as it mixes and spreads across the
Pacific basin by slow bottom currents.

In short, the rate at which any one of the barriers might be breached
could be sufficiently slow as to assure isolation. However it is these
rates, taken together, that will be assessed in relation to each important isotope in the wastes.

We have made some very preliminary and conservative estimates about
rates of breachment that apply to disposal into central North Pacific red clay sediment. They are listed in Fig. 5.1.

FIG 5.1 ESTIMATED BREACHMENT RATES

Barrier	Years Before Breachment/Release
1. Waste form and canister	10^2 to 10^3
2. Sediment (for 100 m burial)	10^6 (pure diffusion only) 10^{13} (sorption + diffusion)
3. Ocean water	10^2 to 10^3

CONCLUSIONS

1. Undersea regions away from active plate boundaries and the productive edges of major surface current gyres offer some conceptual
promise for waste disposal because of their geologic stability and
comparatively low biologic productivity.

2. Trenches are some of the most dynamic and least predictable
regions on the planet and thus are unacceptable as repository sites.

3. The "host medium" or geologic formation remains the single most
important barrier to the release of radioactive material to the biosphere and the two most important retention aspects of the medium are
high sorption capability and low permeability.

4. Another desirable characteristic of the medium is that it behave
plastically or as a visco-elastic medium to enhance its integrity if
disturbed. Such plasticity could also induce self-healing of an
emplacement hole.

Dr. Charles D. Hollister

5. It is necessary to understand the medium's behavior under a variety of thermal fields in order to predict its response as a function of temperature and time.

6. The submarine geologic formation that appears to best satisfy our criteria is abyssal "red" clay which covers approximately 1/3 of the sea floor.

7. The most important constraint for repository siting is environmental predictability over the medium time span needed for waste isolation.

8. Regions likely to be adversely affected by another ice age should be excluded.

9. All regions containing known or suspected concentrations of food, minerals or hydrocarbons should be excluded.

10. From the point of future hydrocarbon recovery entire continental margins including deep-sea fans, cones and aprons should also be excluded. In addition, regions below areas of high biological productivity (such as upwelling areas) which support major fisheries and manganese nodule fields rich in nickel and copper are excluded also.

11. All available data suggest that shallow ocean crustal rock is not monolithic and thus we suggest that emplacement of wastes in the crustal rock, at least at shallow depths, would not be prudent.

12. Potential hazards to man and threats to other ocean uses, as assessed prior to dumping, remain the key factors in determining what should be considered unacceptable "pollution of the marine environment."

REFERENCES

1. Bishop, W.P. and Hollister, C.D., Seabed Disposal-- Where to
Look (1974). Nuclear Technology, 24:425-443.

2. Heezen, B.C. and Hollister, C.D., The Face of the Deep--An Illustrated Natural History of the Deep Sea Floor, (1971). Oxford
University Press, N.Y., 659 pp.

3. Hollister, C.D., Silva, A.J. and Driscoll, A.H. A Giant Piston
Corer (1973). Ocean Engineering, 2(4):159-168.

4. International Atomic Energy Agency, Information Circular 205/Add.
1/Rev. 1, (Aug. 1978). "Convention on the Prevention of Marine
Pollution by Dumping of Wastes and Other Matter," 26 pp.

SUGGESTED READINGS

1. Release Pathways for Deep Seabed Disposal of Radioactive Wastes,
SAND 74-5854 (1975). Presented at the International Symposium on
Radiological Releases from Nuclear Facilities into Aquatic Environ-
ments, Otanieni, Finland. Anderson, D.R., Bishop, W.P., Bowen, V.T.,
Branne, J., Detry, R., Ewart, T., Hayes, D., Heath, G.R., Hessler,
R., Hollister, C.D., Kreil, K., McGowen, J., Rohde, R., Schimmel,
W., Silva, A.J., Smyrl, W., Taft, B. and Talbert, D.

2. Nuclear Wastes Beneath the Deep Sea Floor, SAND 74-5855, 1975.
Presented at Third International Ocean Development Conference, Tokyo.
Bishop, W.P. and Hollister, C.D.

3. The Seabed Option, 1977. Oceanus, 20:18-25, Hollister, C. D.

4. Radioactive Waste Disposal in the Oceans, The Ocean Yearbook,
1977. Frosch, R.A., Hollister, C.D. and Deese, D.A., pp. 340-348.

5. Nuclear Power and Radioactive Waste, 1978, D.C. Heath & Co.,
Lexington, MA, Deese, D.A., 206 pp.

6. Cenozoic Sedimentation in the Central North Pacific. Science.
Corliss, B.H. and Hollister, C.D., in press.

SUB-SEABED DISPOSAL FROM A NATIONAL WASTE MANAGEMENT
PERSPECTIVE

D. Glen Boyer
U.S. Department of Energy
Washington, D.C.

What I thought I would like to do is put into context the Sub-seabed
Disposal Program in relationship to the rest of the Department
of Energy's Radioactive Waste Disposal Program that we call the
National Waste Terminal Storage (NWTS) Program. The overall objective of the NWTS program is to identify technologies that will provide a high degree of assurance that existing and future radioactive
wastes can be isolated from the biosphere in a safe, and environmentally acceptable manner, and to provide the necessary facilities to
achieve that isolation. The major thrust of the Department of Energy's National Waste Terminal Storage Program is towards the isolation
of waste within stable geologic formations, within the continental
United States, at depths reachable by conventional mining methods.
The geological formations that are of primary interest are the rock
salts, granites, shales, and volcanic tufts. A prime alternative
is the sub-seabed alternative. A systemic assessment (as Dr. Hollister has indicated), has shown that some of the very deep-sea sediments have some very attractive characteristics, and are considered
to be some of the more stable formations on the earth. The Sub-seabed Disposal Program is investigating the feasibility of disposal
within these sediments. The near-term objectives are to assess the
technical and environmental feasibility of the concept of engineered
emplacement of appropriately solidified and packaged high level
wastes within the stable deep sea sediments; and to develop and maintain the capability to assess other nations' ocean disposal programs.
We feel this is a significant part of our national responsibilities.

Over the last year there has been a number of studies on alternative
concepts, and what we as a nation should do to manage radioactive
wastes. Most of you are aware that the Inter-agency Review Group
(IRG) on Nuclear Waste Management made its report to the President.
That report was submitted in March, 1979, and while the President has
yet to issue a statement on the waste management strategies, such a
statement is expected shortly. The following key characteriestics
of that strategy are based on the recommendations of the Inter-agency
Review Group and the Department of Energy is following these recommendations.

These strategies include:

1. Mine depositories in geological formations, will be the first
approach for disposal in the United States;

2. Technology is adequate to proceed with geological evaluations
for site assessment;

3. A broad range of geological media and environment should be
explored;

4. Deep ocean sediments should be evaluated as a potential alternative;

5. Technical conservatism should be used for the repository;

6. A systematic approach should be used to select a geological and environmentally acceptable repository; and away from reactor storage of spent fuel can play an important part in a flexible program, leading to the timing of selection of repositories.

The Department of Energy cannot unilaterally decide on and implement a course of action. The Energy Reorganization Act of 1974 requires that the Nuclear Regulatory Commission review and license commercial waste repositories and the National Environmental Policy Act of 1969 requires that environmental impact statements be prepared and become part of the public, state, interest groups, and Federal agency participation in the decision-making process. In April of this year the draft generic Environmental Impact Statement on the Management of Commercially Generated Radioactive Wastes was issued. A series of public hearings aired that document, and the commments received are currently being prepared to issue that report in its final form.

In addition to addressing the major thrust of mine repositories, it includes 10 different alternatives, one of which was the sub-seabed disposal option. Within the National Waste Terminal Storage Program are four major sub-programs each investigating isolation systems centered on different host media and specific related technologies. Under the direction of the Office of Nuclear Waste Isolation of the Battelle Memorial Institute, thus far has been concentrating on evaluation of rock salts, and are beginning to expand their work into areas of granites and shales. Under the direction of the Richland Operations Office basalts are being evaluated in the Columbia Basin region while under the Nevada Operations Office's direction, various geological formations on the DOE's Nevada Test Site are being considered. The Sub-seabed Program is under investigation by the Sandia Laboratory which is under contract to the Department of Energy.

As Dr. Hollister indicated, one of the main objectives of the sub-seabed program is to discuss openly, and to evaluate, the technical advantages and disadvantages of the concept. It is recognized that the ocean disposal option or the sub-seabed disposal option involves international participation. An International Seabed Working Group under the auspices of the Nuclear Energy Agency of the Organization of Economic Cooperation and Development or OECD, provides the forum for discussion of international developments, coordination or research, use of research ship time, exchange of information, and provides an opportunity to review international policies and issues that should and must be discussed if this concept is to be considered a viable option.

We have recently issued a multi-year Sub-seabed Program Plan. We would invite you to obtain a copy and to provide us with any observations or comments.

From the background of this forum today, concerns over legal and institutional requirements need to be assessed. We feel that on a national basis the sub-seabed option needs to be kept open, that we might as a nation and as a member of the world community, evaluate these options in an atmosphere where we can avoid prejudice and at the same time develop the technology and the scientific base to

arrive at an appropriate solution for the problem of radioactive
waste management.

Mr. Dyer outlined (I think appropriately) the background and history
of the present regulations on dumping. From the standpoint of high-
level radioactive wastes emplantation, I would like to make the point
that the U.S. current policies, and the current international laws
all have a background of dumping. That is, a concept of dumping
materials into ocean surface waters and letting it settle to the
ocean bottom. Even the cover of the CEQ 1970 national policy statement on ocean dumping leaves the impression of material which is
dumped perculating from the surface on down to the ocean bottom.
That there is in fact a difference between dumping and the concept
of engineered emplaced material within the geological formations of
the ocean floor.

Within the sub-seabed disposal program it is expected that there
needs to be and should be extensive discussion on the legal and
institutional aspects. One of the sub-objectives of the sub-seabed
program is to identify the requirements of future legislation or
regulations. Within the multi-year program plan is the planned
activity to understand the legal and institutional issues, to develop
a scenario for international management, to develop a domestic and
international discussion on regulations, and to develop the social
and political framework for a basis for properly controlling the
international aspect of radioactive waste disposal.

AN ASSESSMENT OF THE SUB-SEABED DISPOSAL CONCEPT

William Bernard
Office of Technology Assessment
Washington, D.C.

I would first like to say that I appreciate the opportunity to be
here today. As a geologist and oceanographer, I have been working at
the Office of Technology Assessment (OTA) of the U.S. Congress now
for the last year-and-a-half on radioactive waste management problems
and issues. My comments today represent my personal opinion and
should not be construed as an OTA position. the OTA is now is the
midst of its study and our final report is expected in June or in
the later part of the summer.

As you can see from the first presentation on sub-seabed disposal of
high-level radioactive waste, Dr. Hollister is certainly enthusiastic
about this disposal option. As he pointed out, the scientists on the
project have not yet found anything wrong with it. But he also mentioned that he is cautiously optimistic -- I think we ought to take
note of that. I am also enthusiastic about the sub-seabed disposal
concept. However, I want to emphasize at this point that it is
only a concept and should not yet be viewed as a disposal alternative.
It does appear to have a great deal of potential and I think that we
should continue studying the option in the hopes that it may prove
to be a viable technology in the future.

Whereas Dr. Hollister focused primarily on the positive elements of
the concept, I would like to mention some of the problems and issues
that still need to be addressed. Before doing this I would like to
say that, in my opinion, the seabed program is probably one of the
best-managed government programs that I have reviewed. Although
the funding level is not high, the scientists and project managers
are doing an excellent job; it is a well-structured, scientifically
founded program. Because of low program visibility resulting from
low funding, I think they have had an opportunity to develop the
program on a firm scientific foundation.

As Dr. Hollister pointed out, the concept is very attractive. The
seabed sediment beneath the mid-plate gyres is very stable, predictable over long periods of time, and uniform over hundreds of thousands of square miles. Up to this point most of the program efforts
have focused on the resolution of the scientific uncertainties associated with the seabed disposal concept. The scientists at Sandia
Laboratories, DOE's prime contractor on the study, plan to look at
alternative methods for implantation and transportation of the waste
in latter phases of the study. There are still several technical
uncertainties about the actual mechanism(s) for implanting the waste.

As Mr. Boyer pointed out, much of DOE's waste disposal program is now
focusing on the development of mined, geologic repositories. It is
rumored that the President's policy statement on waste management
may adopt Strategy III of the Interagency Review Group (IRG) report,
an approach that is also consistent with the present thinking of the
Nuclear Regulatory Commission (NRC). If that is the case, DOE will
probably be required to investigate several geologic sites, before
submitting three to five to NRC for licensing. Site selection would
probably be made by 1985 with operation of one or more repositories
perhaps in 1995. Not much has been said today about the availability
of the sub-seabed disposal option, but it is my understanding that
it could be a viable alternative in the year 2000. So as you see
there may be convergence on both geologic and the seabed options at
approximately the same time.

The seabed option also has another advantage in that the seabed has
gone through a regional screening process and extensive site exploration. Plans for insitu testing are now underway at one site that
appears to have a great deal of potential. In the case of geologic
repositories we have gone through a regional screening process in
the U.S. We are still in the process of detailed site screening and
surface testing at this point; however, no site has been selected
for repository development. Site selection may be postponed until
1985, depending on the Presidnet's policy statement. So, I think,
in many respects, we have progressed farther on the seabed option
than we have in our efforts to find a mined geologic repository.

In addition, the geologic disposal option is perhaps somewhat complicated by a new emphasis on waste form research. In the past many
people involved in waste disposal had assumed that the geology would
be the primary barrier for containing the waste. However, they have
recently decided that perhaps more emphasis should be placed on the
ability of the waste form to contain radionuclides. So, in addition
to looking at geologic sites, there needs to be a great deal more research and development focused on the different waste forms that
would be considered for geologic disposal. If the sediments on the

seabed are as good an "isolator" as many geological oceanographers
feel, the primary barrier for sub-seabed disposal would be the sediments rather than the waste form.

There has also been a great deal of discussion about retrievability.
The EPA criteria and NRC regulations may demand retrievability for
periods of up to fifty years. Unfortunately, there is a great deal
of controversy about what retrievability means. I agree with
Dr. Hollister that we can find anything on the ocean floor and bring
it back. The technology exists today to do this. The Glomar Explor-
er probably has the capability for retrieving canisters off the ocean
bottom or from within the ocean sediment; however, actual retrieval
of buried waste still needs to be demonstrated. In addition, retrieving penetrometers from within the ocean sediment will probably
be quite expensive and something that wouldn't be cost effective on
a routine basis.

Dr. Hollister also mentioned some of the questions about the sedi-
ment/waste interaction and the importance of the heat generation of
the waste. If the waste is extremely hot, the sediment in the vicinity of the waste would become warmer and less dense than the surrounding (cooler) sediment. With this situation there may be a ten-
dency for the lower density sediment to rise to the sediment surface
and "burp" (or break through the sediment surface). If this pheno-
menon does occur with hot waste, the seabed option may not be a
viable technology for disposing of young waste, but it may be a useful technology for disposing of older, or cooler waste. This, or
course, would necessitate either additional storage time or dilution
of the waste. If we wished to dilute the waste, the volume would be
increased and some sort of spent fuel processing would be involved.
On the other hand, cooling the waste in a storage facility may increase the proliferation risk, simply because the waste would be
cooler and simpler to handle.

If we consider the seabed as a viable disposal option, we must also
evaluate other technologies, such as storage and transportation, to
determine their role in a waste management system. Some type of institutional control for controlling and monitoring the disposal
activities may also be necessary. One of the reasons why we can't
find some of the cannisters of low-level waste at disposal sites is
simply because they were probably dumped someplace else. I think
that it is critical that we keep these types of things in mind.

The cost figures for ocean disposal are also still up in the air. I
think that for both seabed and land disposal options, cost should
probably not be a primary consideration in evaluating disposal
options, but I certainly think that it deserves some attention.

There are also many uncertainties associated with the models that
are being developed to predict the fate of the radionuclides from
low- and high-level waste, for both land and seabed disposal options.
So when making a judgment or "best guess" about the viability of a
technology and the fate of radionuclides, we have to consider the
degree of uncertainty that may be associated with our predictions.

There are also legal, institutional, and political questions that
need to be resolved for seabed disposal. Obviously, if the seabed
becomes a viable disposal option, it would provide a "solution" not

only for the United States, but also for many of the other countries
in the world who are generating high-level waste. There have been
many problems in locating geologic repositories in the U.S. simply
because people don't want it "in their backyard." Although fish
don't vote, there are many lesser-developed countries who are as
concerned as we are about the ability of the seabed to contain the
high-level wastes. I think that it is mandatory that we provide
them with a rigorous evaluation of the concept, especially since many
lesser-developed countries are not generating high-level radioactive
waste.

Over the past few years, the Federal government has gained a lot of
experience with our states in its attempts to find disposal sites
for radioactive wastes. Although we have had many unfortunate experiences in dealing with the states, we can draw on those past
experiences in presenting the seabed option to the rest of the world.
It is my personal feeling that the United States should not unilaterally advocate the seabed option, but instead the option should be
broached to the international community through the International
Seabed Working Group.

As I pointed out earlier, there are many technical and non-technical
uncertainties that must be resolved. I want to emphasize again that
the seabed disposal option is not a "quick fix." Because of the
emphasis on finding a disposal option as quickly as possible, there
has been too much emphasis in the past on finding a "quick fix."
The "quick fix" in the 1960's was disposal in salt; there now seem
to be many questions about the disposal of high-level waste in salt.
In the 1970's the retrievable surface storage facility appeared
to provide a solution; that concept was abandoned in 1975 because it
addressed only storage rather than disposal. Over the last three
years, there has been a shift in emphasis away from containment by
geologic formations, with more emphasis placed on the development
of waste forms and cannisters for containing waste. Many people see
the technologies of a waste management system should be considered
when developing waste management strategies.

Many people have questioned whether the seabed program could be
accelerated. Perhaps it could. But I think we should accelerate
the program only as fast as the scientists and engineers can handle
it. I think to accelerate the program faster could lead to mistakes
which are costly in terms of time, money, and credibility. And I
think that the credibility problem is one that we really ought to
seriously consider. Our experience over the last 20 years in developing waste disposal technologies has shown that in many cases
"haste does make waste."

QUESTION AND ANSWER PERIOD

MR. ROOSEVELT: Thank you very much, gentlemen. I would like to
limit our question period to about 12 minutes. And I direct that
word of caution not only to the audience, but also to the panelists.
I was myself intrigued by Dr. Hollister's reference to tough questions being raised by leading oceanographers. I think Bill has
raised some of those questions. I would like to see if you could
refer to some more of those that have not been raised and I think
that one of the transitional questions that is of critical importance

which Bill has also raised, deals with whether or not our political
constituency questions are going to be deciding the issued at hand
rather than a technical, legal, policy or rationally-oriented process. And that leads us into the afternoon program which deals with
legal and policy considerations. But let's handle the questions
now for this particular session.

QUESTION: What is the area under study for possible sub-seabed
disposal?

DR. HOLLISTER: We are basically looking at the region of the central north Pacific, Mid-Plate, mid-gyre region. It is essentially a
region between 10 and 40 degrees north latitude and at least 200
miles from coastal region. That is just one of the areas that we
are looking at. Of course, in the European community we are looking
at areas, off of Cape Verdes, south of the Azores, and a region
somewhere north of the abyssal plain, near Bermuda (on the Bermuda
rise). The areas that are being studied are about 100 miles on a
side. These areas have been identified from historical data, core
samples, (whatever we have in the laboratory) and from seismic pro-
files, we are trying to determine if these areas are worthy of
further consideration. Because we don't have much money, we usually
have an expedition to an area of from two to three weeks duration per
year, and then it takes about another year to assess all the data.
The regions are vast and we do have a great deal of knowledge so
we started choosing the study sites based on historical data.

QUESTION: I wonder if you could specify some of the tough questions
that oceanographers are posing?

DR. HOLLISTER: I think that one of the toughest that we are posing
to each other is how do you predict the behavior of a water saturated
viscoelastic clay when you emplace hot canisters into it? We start
by making a mathematical model for say; dispersion rates through
relatively impermeable clay that predicts a certain behavior and
then we try to verify it. Perhaps first in a laboratory and then on
the seafloor itself.

Like in any complex model verification effort, the first attempt is
generally off, so we go back to the laboratory and think of another
model and try it again.   Often we find that we do not have the
right property values so we go back and measure different parameters,
or remeasure the faulty ones before proceeding to a verification
effort. The major effort, of course, is to answer the question:
is the sediment a primary barrier to release after the waste has been
emplaced?

If it turns out that the sediment is a questionable barrier, then I
think the whole effort should be reevaluated. At this stage of our
research we are not taking much credit for the waste form or the
canister as a barrier except during the maximum heat period of about
300 to 500 years.

There very well may be several unknowns within our dispersion-rate
model that we will never be able to answer. We will come back to
the public and say, "these are the unknowns, do you accept the risk
or don't you?" You will have to make the final decision to accept or
reject by weighing the alternatives of the sub-seabed disposal option.

QUESTION: How do you obtain samples from the deep sea floor?

DR. HOLLISTER: By lowering a core tube to the bottom or by drilling with a rotary coring device. So far, the Deep Sea Drilling Project has drilled over 500 holes into the seafloor and since we started taking sea floor samples about 100 years ago we have taken countless core and surface samples of the deep sea sediments. We do a lot of carbon, nitrogen and oxygen determinations on the sediments in our research laboratories.

The types of sediment we are looking at for the sub-seabed disposal effort has an organic carbon content of about 0.1 percent or less. It is highly oxidized ultra-fine grained and impermeable clay with very low rates of accumulation; like 0.1 milimeter per thousand years. This oxidized clay has the highest ion exchange capacity of any geological formation that we have studied so far. I guess that we have tens of thousands of samples of this clay formation in our large core repositories at major oceanographic institutions like Woods Hole Oceanographic Institution.

QUESTION: How much is it going to cost and who is going to pay for it?

DR. HOLLISTER: Preliminary cost estimates suggest that the cost of an actual working mined repository in the continental U.S. with all of its safe guards and monitoring systems in operation would be about the same as an international repository in subseabed formations.

QUESTION: Specifically, has the Department of Energy run any tabs on the cost?

DR. HOLLISTER: They all run tabs all the time. That is all they seem to do is run tabs.

MR. BOYER: The current expenditures from 1974 to date on assessing the seabed disposal concept is approximately $10 million.

QUESTION: What I meant was, what is the estimated cost for disposal of radioactive waste.

MR. BOYER: That number has not been generated for the seabed disposal option. If I remember correctly, the total cost estimated for disposal of the mine repository is in the area of $20 to $25 billion. That is for handling that waste which is presently expected to be generated by commercial operations through the end of the century.

QUESTION: Does that include the research and development cost?

MR. BOYER: No, that is a total cost. And the concept is that for those wastes that are generated commercially, the costs would be back-charged to the user. Of course, there are those wastes which have been generated through the defense program which is the government's responsibility. Yours and mine.

DR. BARNARD: I could argue with those cost figures. I think they are quite optimistic.

DR. HOLLISTER: Until we know what we are doing, I think that is a ridiculous question to ask.

QUESTION: It seems to me, at least that in the case of oil, the
problems are the transport phase. You could answer the scientific
questions of isolation of waste to everybody's satisfaction perhaps,
but the practical problems of marine transport, weather and so on,
appear to me to need serious attention. I was wondering what DOE
or EPA are doing about that consideration of transport as well as
the monitoring, packaging, and retrievable portions of the program.

MR. BARNARD: One of the models that Dr. Hollister talked about is
that you make an assumption, and then you identify the risks of
those assumptions and you are correct. The assumptions which need
to be assessed from the acceptability include the concept of land
transport, to the port facilities, handling at the port, handling in
the ship, what are the risks of accidents there, then the placement
into the sediment, I think we have much more confidence in what we
know about the sediments. The areas of risk assessment still need
to be done.

QUESTION: Is that actively under consideration and is there a program being considered?

MR. BOYER: Yes, within this sub-seabed multi-year plan, it involves
each of these various components and one of those components is
transportation and the risk analysis.

QUESTION: Could you describe some of the international efforts of
your proposal?

DR. HOLLISTER: Yes, when we started this whole effort the first
thing we did was to talk to our scientific colleagues and ask,
"What do you think?" Out of this group of international "reviewers"
we formed a "Seabed Working Group" which now includes Canada, U.K.,
and France and to a much lesser extent by the Netherlands, Canada
and Japan. We are also keeping the "law of the Sea" folks and the
principal environmental and public interest groups like the Oceanic
Society fully informed of what we are doing.

QUESTION: What is a likely area that would be needed for a sub-
seabed repository?

DR. HOLLISTER: We are looking at study areas in the centers of a
huge stable lithospheric plates, that are about 100 miles on a side
in area. Which if canisters are placed on a hundred meter centers
could accommodate one million canisters.

6. SUB-SEABED DISPOSAL IN THE CONTEXT OF THE LAW OF THE SEA

The Hon. Elliot Richardson
Ambassador, Law of the Sea Negotiations
U.S. Department of State
Washington, D.C.

INTRODUCTION

Ambassador Richardson speaks to the difficulty of weighing issues as
complex as subseabed disposal on a scale of costs and benefits. The
volume of data available to assist in policy formation also compli-
cates the society's process of determining trade-offs within a
framework of tolerable risks.

Despite this difficulty, it is clear we must continue to use the
best available information and consider the environmental impacts of
proposed policies, the ambassador notes. At some point in a policy
debate, he adds, disagreement of opinion stems from a disagreement
on the facts. This kind of controversy is settled by the premium a
society places on expanded consumption and energy as opposed to the
premium placed on preserving the ocean environment. After outlining
a process to resolve these conflicts, Ambassador Richardson places
the subseabed option in the context of current Law of the Sea negotiations.

Two key issues raised by questions focused on the effect of a Law
of the Sea treaty might have on subseabed disposal and also the concerns of Third World countries on the effect nuclear waste disposal
at sea may have on the "common heritage of mankind." (Editor's note).

SUB-SEABED DISPOSAL

It is my distinct pleasure and honor to welcome and introduce to
you Ambassador-at-Large Elliot Richardson, who is the United States
delegate to the ongoing United Nations Law of the Sea negotiations.
Ambassador Richardson. (Mr. Roosevelt)

Thank you very much, Mr. President. It sounds rather suitable to
address a Roosevelt in that way, don't you think?

Ladies and gentlemen, participants in what would seem to be an extremely valuable exchange of views, information and judgment among
many highly gifted and highly informed people. Indeed, as I reflected
on the character of this session I was about to address I came increasingly to wonder what I was doing here. But I am pleased to
greet you on a day which to me is very important because I have now,
as of this week, established for myself a new record in job tenure.

It occurred to me in reflecting upon the theme of this symposium that
there may after all have been more profound significance in the

Fig. 6.0   The Hon. Elliot Richardson

legend of Prometheus than we may have been disposed to believe in
recent decades during which mankind has been riding higher and
higher on a wave of technology. You may recall that Prometheus was
sentenced by Zeus to endure for eternity an extremely painful form
of punishment in return for the sin of having stolen knowledge and
bestowed it upon mankind as a gift. And because of this, Zeus con-
demned him to be chained for all time to a rock where his entrails
were constantly being gnawed by a vulture. Reflection upon the
dilemmas created by knowledge and the capacity to use it can help to
provide perspective on the justice of the sentence imposed by Zeus
for the crime of giving knowledge to mankind. Even as recently as
the early years of my own career in government service there were
times when we thought we could reach a relatively clearcut decision
in a relatively complex situation without having to engage in an extensive analysis of costs and benefits. Now, of course, for better
or worse, we know better. We understand more or less what a system
is--even an ecosystem--and what is implied by the interdependencies
among the elements comprising a system of systems.

I take a system to be a structure such that if you exert an impact in
one place, it will bulge somewhere else, or if you put a strain on it
in one place, the strain will be transmitted throughout the structure.
The subject of deepsea disposal of high-level nuclear wastes could
hardly symbolize more dramatically the kinds of complexities and
trade-offs that mankind is constantly going to be called upon to
cope with--at least to attempt--until, perhaps, the point is some day
reached when we shall have to throw our hands up in despair and
plunge ahead into a future in which the possibility of calculating
all the consequences of some possible impact on the system of systems
in which we live is so discouraging that we give up entirely.

For the moment at least, analysis is still possible and important,
even necessary. There is, to begin with, the enormous set of technical issues antecedent to the problem of the disposal of radioactive wastes that is raised by the question of the pace at which
the world community will generate radioactive wastes. The answer to
that question, of course, will depend in the long term on what is
ultimately decided as to the necessity for reliance on nuclear energy.
The degree of that necessity will in turn be affected by the question
of the disposal of the wastes thereby generated.

This is an obvious point, I am sure, to all of those of you who are
concerned about the subject, and yet it illustrates a threshold
difficulty, a problem of circularity, inherent in trying to strike a
balance among the costs and benefits, at a tolerable risk, of
nuclear power generation. Assuming that it were possible to resolve
that set of problems and thus to develop a projected quantum of
forseeable waste accumulation, we would then come to all the kinds of
issues that you are addressing, and it is fair to say that these are
issues that could not be addressed at all but for the accumulation
over a long period, and especially in recent decades, of knowledge
itself. Take, for instance, the measurement techniques that allow
us to be aware of risks that we could once have ignored either because we could assume that they didn't exist or, even if we conjec-
tured that they might, we could not in any case measure. Take resi-
dues in pesticides or the ability to detect one part per billion of
diethylstilbestrol in beef. I am sure that it did not at first occur
to people that they should worry about the use of very small amounts

of this hormone for fattening purposes. But once we have learned
how to measure parts per billion, we then have to factor these findings into the calculation of tolerable risk. And, as epitomized by
the problem that you are discussing, these are risks not simply of
today or tomorrow but for the very long future.

What is the quantum of a risk on a given scale affecting a given
number of people over an assumed length of time? How do you equate
or reduce to some common scale risks such as nuclear energy that involve potential damage to thousands, perhaps hundreds of thousands,
of people but which are very unlikely to occur, as against much more
certain dangers, that are with us all the time? It has often struck
me that people who support the Delaney Amendment's prohibition of any
possible trace, no matter how small, of a substance that has ever
been shown to produce cancer lesions in rats, even though the quantities ingested by the rats were enormous, are the people who if they
applied that same standard to their daily lives would live in one-
story padded houses without bathrooms and never go outdoors. The
question of what are acceptable risks and how you compare them is,
of course, central to the issue you are dealing with as is the
problem of minimizing the risks, whatever they are.

You then come to the problem of how to encapsulate the radioactive
wastes and of how to propel them into the sediment of the ocean
floor. And the decision of whether or not to use the seabed floor
for the disposal of radioactive nuclear wastes must take into account
not just potential danger to the ecosystem of the deep ocean, but
the potential risks of alternative means of disposal, including the
seepage of radioactive waste through the fissures in the walls of
subterranean salt domes.

All of this underscores a concern that I do not see being adequately
addressed by the United States or the world community generally. The
problems of policy and choice of the kind that you are here to discuss and which I have just been touching on lead at many points to
what are essentially differences of judgment over the fundamental
facts. This is true of almost any complex policy issue in today's
world, whether it is a question of how to increase productivity, how
to combat inflation, or how to deal with the problems of health in-
surance for the elderly. I have acquired over the years a very sensitive ear to the point at which a debate on an issue of policy
turns into a debate over the facts. In such situations, I have
occasionally suggested to the responsible people that they look back
over the record of the discussion with an eye for the points at which
the record would show that a difference of opinion had arisen from a
disagreement about the facts. And I might suggest, Chris, that this
might be worth doing as you review the record of this meeting. In
this way, we can at least identify opportunities for narrowing the
range of policy dispute.

The fundamental choices among values cannot be avoided by any process
of inquiry or by gaining futther knowledge. These will turn, in the
end, upon such things as the relative premium that society places on
the expansion of consumption and energy use as against the premium
it places on preserving an ocean environment capable of supporting,
say, whales and porpoises three generations from now. No scientific
finding can tell this or any other generation what value we should
place upon the quality of life of people who will be born long after

we are gone. But we can at least place a fundamentally difficult issue of choice among competing values within a factual context that is as firm as we can make it. This requires a process whereby the factual issues that have been identified can be subjected to a systematic effort by experts to see how far they can come to agreement on the basis of the available data. This would be not an attempt to force the reconciliation of any views or to compromise differences, but rather to see what can be the subject of valid and accepted findings.

The next step would be to seek agreement on approaches toward finding answers to questions left open by the first step. This would involve formulating the questions and might also include recommendations as to how and under what auspices the additional inquiries should be carried out. And then, because the policymakers might have to make choices before these additional steps can be taken or perhaps because it is recognized that adequate answers to some of these questions may not be available for a very long time, if at all, it would be desirable to be able to tell the policymakers what are the parameters of uncertainty within which their choices will have to be made. And here again we come back to the question of acceptable levels of risk.

Various approaches, as I am sure you know, have been suggested to the problem of how best to obtain consensus on scientific and technical issues. When I was Secretary of Commerce several years ago, the Department sponsored a discussion involving people from many fields on the question of the desirability of creating a Science Court whose function it would be to address this kind of issue. Although I think that there is some merit in the idea, its limitation is that the attempt to determine the extent of agreement on a technical issue or to identify opportunities for further inquiry, and so on, would be cast in an adversary mold. I question the wisdom of doing this, especially since in most instances, there will be a need for a continuing process of seeking further information. But the importance of the idea is that it at least reflects awareness of what in my view is a seriously neglected problem.

In the case of the disposal of high-level radioactive nuclear wastes you have a problem that illustrates the need and the opportunity for this kind of follow-up process. This is why I suggested a moment ago that perhaps it would be worthwhile to look back over the papers or any summaries of this discussion to see to what extent these may point to the need for further inquiry.

So far as the legal framework within which the ultimate policy decisions are carried out is concerned, I think that here for once the lawyers have a considerably easier role to play. I have been looking at the Convention on the Prevention of Marine Pollution by Dumping of Wastes that was adopted in London in 1972. As you know, it went into effect in 1975 and had, at last count, been ratified by about 40 countries. I have also looked at the provisions of the Law of the Sea texts we have been negotiating in New York and Geneva over the last several years. The environmental provisions of that treaty are in more nearly final form than perhaps any others and so can be regarded as unlikely to change very much, if at all. Basically, the approach taken in the London Dumping Convention as well as in the Law of the Sea Treaty is an approach which "bucks" the ball back to the scientists and technologists. No attempt was made by the inter-

national community (least of all by the lawyers) to determine what
the applicable standards should be. These are the standards that
will be established in the first instance by the IAEA in defining
high-level radioactive wastes unsuitable for dumping at sea on the
one hand and those other matters not so designated.

As you know, the London Dumping Convention called for a complete prohibition on the dumping of high-level nuclear waste and makes the
disposal of lower-level waste subject to a permit system based on internationally accepted criteria for disposal. The responsibility for
defining high-level radioactive waste and for recommendations for the
terms for ocean dumping of other radioactive waste has been assigned
to the International Atomic Energy Agency in Vienna. Among the
standards to be established by IAEA is the depth of the water at
which, if at all, radioactive disposal will be allowed. In general,
the result is to defer the technical issues to the technical community, but to back up these standards with an enforcement system that
can make them effective.

This is where the Law of the Sea treaty comes in. The negotiating
text calls upon member states to establish national laws and regulations to prevent, reduce, and control pollution of the marine environment from dumping and to join in establishing global and regional rules, standards, and procedures through the competent international organizations. Dumping within the territorial sea and the
exclusive economic zone or on the continental shelf would require
the express prior approval of the coastal state. The standards established by the IAEA with respect to radioactive waste disposal
would be enforced in the territorial sea, the exclusive economic zone,
and the continental shelf by the coastal state, and by any state
with regard to acts of loading of waste or other matter within its
own territory and its off-shore terminals. In the case of dumping at
sea beyond the reach of coastal-state jurisdiction, responsibility
will rest with the vessel and with the flag state, but any country
concerned with the monitoring or observance of these standards could
request action by the flag state. There will also be an opportunity
to recover on account of any damage done, but this, in the case of
radioactive waste disposal, may be difficult to establish. And there
would be the opportunity, as in the case of the vessel-source pollution standards developed by the Intergovernmental Maritime Consultative Organization (IMCO), progressively to up-date the applicable
standards in the light of evolving technology.

It would thus appear that the framework for the adoption and enforcement of standards will be reasonably adequate and reasonably firmly
in place if and when a Law of the Sea treaty comes into force as the
most universal system within which accepted standards can be enforced.

And here it may be useful to say a little more about the Law of the
Sea Conference. It is a Conference in which all countries are represented, about 160 in all, eight more than the entire membership of
the United Nations. In one way or another all of these countries
have identifiable interests in some aspect of the whole range of
problems presented by man's concern with and uses of the oceans.
The inevitable result is that the drafting of a comprehensive code of
law is an exercise in balancing politico-military interests in free-
dom of navigation and overflight with protection of the marine environment and coastal-state jurisdiction over economic resources with

the conduct of marine scientific research, and so on. All of this is part of a comprehensive package. If, therefore, the conference can succeed, and it is not too far from doing so, there will be greater incentives to ratification by a large number of countries than would ver attach to the London Convention on Ocean Dumping by itself or an IMCO convention on vessel source pollution by itself. When you add that the Law of the Sea Treaty would make enforceable what it refers to at many points as "generally accepted" standards, you can see that a large number of ratifications of the Law of the Sea Treaty would have the effect of extending enormously the impact of the specialized conventions because these are the source of the "generally accepted" standards. An example is the IMCO standards on the rate of discharge by a tanker of the oil mixed with its ballast as it proceeds somewhere out at sea. These discharge standards at present are only enforceable against parties to the pertinent IMCO Conventions, but under the Law of the Sea Treaty they would be drawn into the treaty as generally accepted.

So, by the same token, what is done to establish the applicable definitions for high-level radioactive wastes will become binding against any country that ratifies the Law of the Sea Treaty, even though it is not a party to the London Convention. It is for this reason, that the Law of the Sea Treaty can thus accomplish in one long stride a strengthening of the protection of marine environment that would otherwise take decades. It may similarly assist in making effective the standards that are applied to the enormously difficult problem that you are all here to discuss. And since I have no competence to offer any advice on how you are to deal directly with that question, I can only say to you in conclusion that as a citizen I am grateful to you for undertaking these deliberations and I am grateful to the Oceanic Society, the Georgetown Law Center, and the Center for Law and Public Policy for having sponsored this meeting. Thank you very much.

QUESTIONS AND ANSWERS

MR. ROOSEVELT: The Ambassador will entertain any questions if there are any from the audience. You covered your topic very clearly I see.

QUESTION: Does the Law of the Sea Treaty cover burial beneath the bottom of the sea? And if so, and if radioactive wastes are 30 meters down below the seabed, would that be affected by the Law of the Sea Treaty?

AMBASSADOR RICHARDSON: Well, I understand that to be a question as to the applicability of the London Convention itself, in the first instance. And the question as to whether or not it is under the London Convention turns on the predictable effect on the environment in the area of the dumpsite. I think that if you could find that there was a risk to the ecosystem of that part of the bottom, it would not only come under the London Convention but the very same findings would also bring it under the Law of the Sea Treaty. Conversely, I think that if you could find that the waste was buried deeply enough so that there was in fact no effect outside the layer or above the layer of sediment in which the waste was buried in some capsule, and if you never found any such effect, then I think it

would be taken outside of the scope of both international agreements
--and probably should be. The definition of dumping is nearly the
same in both the London Convention and the Law of the Sea text but
it is fairly arguable that they do not cover emplacement beneath
the seabed.

The question that then arises -- and I don't know the answer -- is
"Well, if you start out with the assumption that the capsule is
buried deep enough and is enough shielded by the layer of sediment,
how will you know whether or not that remains true?" And given the
possibility of shifts in the seabottom and the opening up of fissures,
and so on, under what legal auspices or jurisdiction is any monitoring responsibility placed? I don't know the answers to those questions as they would be affected by the Law of the Sea Treaty, and
perhaps they ought to be considered in establishing a really adequate
long-term regime for dealing with this subject.

QUESTION: To what degree do the non-nuclear Third World nations express concern over nuclear nations using the common heritage of mankind, the seabed, for disposal of wastes?

AMBASSADOR RICHARDSON: I haven't heard the problem raised in the
Law of the Sea conference itself. There has been almost no discussion of nuclear waste disposal, perhaps because of a general awareness that the subject is being pursued in other forums. So far as
I've hear it addressed at all, it is only as a passing reflection of
the broader concern of developing countries that they are being
asked now to accept environmental protection standards that were
adopted and became enforceable in industrial countries only after the
latter had reached the stage of advanced industrial development. And
there is, of course, an underlying feeling that the developing countries ought to be allowed to catch up; that every country is entitled
to cause a certain amount of global pollution before it has to accept universal standards. But that, of course, is not an attitude
that stands up very well under any real scrutiny. It is just an
undertone that is part of an overall feeling that the industrial
countries that got there first did things their own way without con-
sulting us, and now that they've got theirs, they are asking us to
observe rules that they neven even thought about before.

QUESTION: Do you see any impact of the Russian-Afganistan situation
and the reaction of the world community to Soviet aggression in
Afganistan in the negotiations of the Law of the Sea?

AMBASSADOR RICHARDSON: I don't know yet what problems there might
be. The Law of the Sea Conference reconvenes in the week of Febru-
ary 25, 1980. Ordinarily we would have considerable and detailed
discussions with the Soviet delegation among others at the very early
stages of any new session of the conference. We had talked to the
Soviet delegation and others before the beginning of this year and
before the business in Afghanistan, but I have had no information or
no indication since. Where the United States is concerned, we must
proceed on the basis that the Soviet Union and others, including
ourselves, are approaching the issues on the merits and in the light
of our national interests and that there simply is no room in the
negotations as complex and many-sided as this for the introduction
of extraneous political considerations of any kind. It is interesting that in the three years that I have been with the conference,

and I suppose this was true before that, I have never seen introduced extraneous political considerations, except where they could
be directly related to a particular problem that did belong in the
conference, such as who should be allowed to sign the treaty. As
you can imagine, the Arab countries want the Palestine Liberation
Organization to be allowed to sign. But I only know one instance
in which bilateral relations, even though seriously strained, have
affected the negotiations at all. And the reason, as I just
suggested, was because they are so complicated nobody can gain any
margin of advantage that can be expended for extraneous political
purposes, and no one can afford to accept sacrifices of significant
long-term interests in the outcome of the Law of the Sea negotiations
in order to gain some external political advantage. So, I would
guess that the probable result is that Afghanistan will have very
little effect on the Conference. The main problem may turn out to
be that the Soviets may not be able to get their plane into New
York.

QUESTION: What impact do you expect to see in the Law of the Sea Conference as a consequence of the discussion on the nuclear fuel cycle?

AMBASSADOR RICHARDSON: It is not obvious to me what effect it could
have. Of course, these negotiations will affect the pace and
development of nuclear fuel production and fuel processing and waste
disposal, and so on. But I think the results would be assimilated
into Law of the Sea in the manner I described earlier. I don't know
of any force internal to the Law of the Sea Conference negotiations
that would be likely to try to influence the outcome of the nuclear
fuel cycle issues.

QUESTION: Is it true in the light of what you have said that the Law
of the Sea Conference does not contemplate any action that would prevent the disposal of radioactive wastes in the seabed?

AMBASSADOR RICHARDSON: The answer is yes, you are correct. The
Law of the Sea treaty will not, for example, contain a prohibition
against the disposal of high-level radioactive wastes, although the
effect will be to assimilate the prohibition in the London Convention,
leaving open as that does, the question of definition. But in so
many words it won't do so. If the signatories of the London Convention as the standard-setting body, someday decide that high-level
radioactive wastes have a benign influence through the encouragement
of more rapid mutations of edible bottom-dwelling species, and,
therefore that the prohibition was a bad thing and should be repudia-
ted, then its assimilation into the Law of the Sea, through the Law
of the Sea Treaty, would also lapse.

QUESTION: Aside from what you have said about assimilation, wouldn't
you think that the International Seabed Authority, established to
manage the exploitation of deepsea bed resources and seabed mining,
would itself assume jurisdiction to control the disposition of radioactive wastes?

AMBASSADOR RICHARDSON: Well, that is a good question, and I haven't
thought of it. But I don't believe the authority could do that because its jurisdiction would be limited to the control and management
of the exploration of deep seabed resources. The exploitable resources of the deep seabed would come within its jurisdiction as

distinguished from the resources of the water column above the seabed. The International Seabed Authority would have ample authority
to establish regulations for the protection of the marine environment as the consequences of the exploitation of any seabed resource.
The Authority could apply and enforce these standards as against a
company involved in deep seabed mining, even to the point of forcing
the company to stop a mining operation because of new evidence that
its activity was significantly destructive of the marine environment.

But of course it doesn't follow that an authority set up only to
manage the exploitation of seabed resources and to protect the marine
environment from the consequence of such exploitation could protect
the seabed generally.

QUESTION: Is there any particular minimum depth above which the disposal of radioactive waste would not be allowed?

AMBASSADOR RICHARDSON: The United States has recommended, and I
believe the IAEA is in the process of adopting, a prohibition against
dumping in water less than 4,000 meters in depth.   The IAEA has also
recommended packaging requirements designed to ensure that the
contents do not leak during the descent to the seabed.   They have
also proposed a standard for the specific gravity of the package.

QUESTION: Inaudible for transcription.

AMBASSADOR RICHARDSON: The question points to the distinction between the agreement in London in 1978 on the establishment of the
4,000 meter standard as a recommendation to member governments and
the steps required to actually bring it into force.

QUESTION: Inaudible for transcription.

AMBASSADOR RICHARDSON: I'm not sure you all heard that, but it concerns the distinction between what happens under Annex 1 and Annex 2,
which deals with wastes other than high-level wastes.   The IAEA has
the power to establish guidelines under Annex 2 for other wastes,
and it is under that power that the 4,000 meter standard has been
recommended, or established.

Well, we have moved by a gradual and perhaps imperceptible process
from the generalizations with which I began and with which I feel
comfortable to the real subject matter of your meeting, so I think
it is suitable that we consider this to have been the last question.
I thank you all very much for your participation and your interest.

Fig. 7.0 Clifton Curtis

Fig. 7.1 David Deese

7. NATIONAL AND INTERNATIONAL POLICIES AND PRACTICES

INTRODUCTION

Deliberations on the "ocean alternative" must also be cast in the context of international policies and practices. Dr. David Deese approaches this discussion first by focusing on the social and scientific problems associated with nuclear waste management and second by addressing issues specific to the sub-seabed disposal option.

After studying the radioactive waste management decision making process in 10 countries, Dr. Deese reports six characteristics can be used as a basis for comparison and to show contrasts. These characteristics can be used as a basis for comparison and to show contrasts. These characteristics range from a technological bias to a fragmentation of government structure and regulatory overloading. Public participation is another characteristic noted by Dr. Deese.

Dr. Deese surveys some of the important policy questions facing the United States sub-seabed disposal program, noting the influence the United States exerts in this field among other nations.

Clifton Curtis addresses United States policy trends affecting ocean disposal of nuclear waste. After setting the dual goals of enhancing environmental protection of the oceans and increasing opportunities for public participation, Mr. Curtis presents a detailed review of current American policy trends.

During this presentation, Mr. Curtis points out some of the difficulties confronting efforts to develop a comprehensive and consistent approach to nuclear waste management. National policies which shape our protection and use of vital marine resources must, Mr. Curtis concludes, be subjected to continuing and vigorous review by the public.

The panel discussion on national and international policy considerations includes an examination of U.S. policy by Edward Mainland of the U.S. Department of State's Office of Environment and Health; a characterization of sub-seabed program as caught between the attitudes of "not in my backyard" and "out of sight, out of mind" by Dr. John Kelly of the Center for the Study of Complex Systems at the University of New Hampshire; a look at United States ocean disposal policy in the context of international accords by Alan Sielen of EPA's Office of International Activities; a concerned view of current nuclear power regulation as oriented toward accommodation to industry pressure rather than holding fast to protect the public health voiced by Arthur Tamplin, Senior Staff Scientist with the Natural Resources Defense Council; and a review of Congressional concerns focused on the "ocean alternative" as seen by Richard Norling, staff director of the Subcommittee on Oceanography. (Editor's note).

INTERNATIONAL POLICY CONSIDERATIONS IN THE
OCEAN DISPOSAL DEBATE

David A. Deese
Research Fellow
Center for Science and International Affairs
Harvard University
Boston, MA

I would like to start by emphasizing that there are few clear answers
to the questions being raised. The best that I can do after investigating these issues for the past five years is to insure that the
important questions are being asked and that the answers are being
addressed with the appropriate scientific methods.

I will offer two sets of comments. The first deals with the overall
social scientific problems associated with radioactive waste management; the second addresses issues specific to the sub-seabed disposal
study. Dorothy Zinberg and I have just finished a comparative study
of decision making in radioactive waste management. We studied ten
countries internationally and documented the similarities and differences among them in handling the radioactive waste management
problem. It was surprising to discover a high degree of similarity
among countries in mistakes made in radioactive waste management.
Some of the strengths that have been discovered in other countries'
programs and held up as models for the United States don't stand up
under closer scrutiny. The results of this recent study set a general context for decision making which is also the context for the
social science of sub-seabed disposal.

I will also draw on conferences and meetings that have been held in
1979. I helped organize a series of eight workshops on specific
issues in radioactive waste management which were held at the Key-
stone Center for Continuing Education in Colorado. Comparative political and broader international issues were highlighted at a workshop in September. The Aspen Institute sponsored a workshop at
Harvard University in November. An organization in California called
"Resolve" has held informal meetings since December, 1979.

I find six characteristics on which these waste management programs
can be compared and contrasted. The first is technological bias in
decision making. Countries have considered radioactive wastes to be
essentially a technical problem requiring technical managers and
solutions. This has obscured the issues and delayed solutions since
the social, political and institutional dimensions are equally important.

There is an extremely delicate balance to be struck between emphasis
on the physical and the social scientific issues. One must immediately say: "Okay, if we slip too far over towards treating it as
a legal problem, then where are we?" We can easily lose sight of
the basic sciences, which is the foundation for public, congressional,
and regulatory responses. This may in fact be where we are currently
headed. We have found that in at least Germany, Sweden, and the
United States, and perhaps somewhat in Japan and Britain, the scale
seems to be tipping too far. In some cases, governments are beginning to look at political boundaries and legal requirements as

driving this system. It is clearly a fine balance that demands wide
discussion and analysis.

Second, there is a striking lack of national strategies for decision
making and implementation within countries for handling the waste
problem. In the United States, critical inertia was lost during the
months between completion of the Interagency Review Group report
and the Presidential announcement. We are now in the midst of an
election year and not much progress is expected.

One of the most important elements of a national strategy is the
selection process for options for final disposal of high-level radioactive wastes. Outside, unbiased individuals and groups should study
the Department of Energy's program, consider the options that are
being investigated in other countries including sub-seabed disposal,
analyze the budgets for the different disposal options, and determine
the basis for setting priorities among them. People should consider
the weight of available technical evidence and decide for themselves
when the narrowing process should begin to select disposal options.

The third characteristic is fragmentation of government structures.
The United States has swung from one extreme, with one commanding
federal agency and congressional committee, into the other, with over
a dozen agencies and two dozen congressional committees asserting
various degrees of control. There has been some important progress
in this area in the last 18 months, but many problems remain.

Fourth is regulatory overload and paralysis. The plight of the NRC
and the EPA is not endemic to the United States or to the nuclear
power industry. It is a profound and widespread problem. Social
expectations are very high for regulators right now. Ambassador
Richardson referred to enormous growth of new knowledge about numerous environmental threats. The regulatory agencies have limited
budgets; they have personnel limits; they have a clear limit to the
technical data base on which they act; and finally, they face uncer-
tain and overlapping mandates from the other governmental bodies and
unclear signals from the broader public.

The fifth characteristic is complex relations between local, state
and federal governments. This is a serious problem in at least
Japan, the United Kingdom, the United States, and West Germany.

The final characteristic is public participation. Federal countries
are now scrambling to develop strong information flows. Some are
also developing mechanisms which allow public influence over decision
making in radioactive waste management. This is where Sweden is
always held up as the example. Caution is the rule if somebody holds
that example up to you and says "all we have to do is look at what
Sweden did" and follow that same process, because as you get into it
in more detail, this is a red herring in several ways. There are
some interesting aspects in the Swedish process, but there are also
others that are completely irrelevant to the United States.

I would now like to survey some of the important questions for a
sub-seabed disposal program in the United States. First, the legal
situation vis-a-vis sub-seabed disposal, then the political factors,
and then quickly the international dimensions. The key legal requirement in the United States is the Marine Protection, Research and

Sanctuaries Act of 1972. It was passed long before the sub-seabed
disposal program came along. It is absolutely clear that when it was
passed, people did not have in mind a sub-seabed disposal program.
Yet this is by itself no reason to exempt it from coverage by the
1972 act. The question that arises concerns an exclusion in the
Act's definition of dumping. The exclusion covers activities other
than disposal in the submerged land beneath the oceans. When I wrote
a book a few years ago entitled Nuclear Power and Radioactive Waste:
A Seabed Disposal Option?, a number of lawyers told me that by a
twist of legal reasoning, the reverse of that exemption means the
things that are for disposal that go into the submerged lands beneath
the oceans are included in the act. Now that may or may not be
twisting the wording or spirit of the Act. I'll leave that to your
own judgement, and to the lawyers who are experts in statutory
interpretation.

In the regulatory area, I only want to say that there are important
roles in this activity, first in the research stage and second, if
it is even implemented in the 1990's or after, for a number of different agencies. The Department of Energy has lot of de facto regulatory authority since it will carry most of these programs for many
years before they come under direct regulation. There are a lot of
rules that NRC and EPA have to promulgate soon if they are going to
have an important effect on DOE programs. There are special rules
from the Department of Transportation and the Coast Guard if we
ever go ahead with a sub-seabed disposal program.

With respect to U.S. policy, I will just mention that there are important roles for NOAA, the Department of State, the Arms Control and
Disarmament Agency and even the National Security Council. These
were in part defined by the inter-agency review group in its report
in 1979.

I just want to make one final statement on the United States situation. In examining again the Department of Energy and its policy
for high-level radioactive waste disposal, I urge you to apply the
old rule that a budget is a statement of policy. Look at the amount
that is spent on each disposal option and see whether you find a
sound balance.

On the international level, legally, there are several important
treaties and statements that have been made by different countries;
there is some customary international law; and there are a number
of ongoing activities which affect the possibility that a sub-seabed
disposal option could be implemented. The focus is the Ocean Dumping
Convention of 1972 that was signed in London and ratified, as
Ambassador Richardson said, by over 40 countries. The key here again
is how to interpret the definition of "dumping". I would just tell
you what I have written without giving you any clear, legally authoritative commitment. I find that since dumping is disposal of waste
"at sea", we have to either define "at sea" as any activity that is
conducted at sea (meaning from a ship or an airplane, or whatever)
or as any disposal of material into the oceans. These two different
interpretations have different results in their application to sub-
seabed disposal. It is clear that when the Convention was signed in
1972, no one had in mind sub-seabed disposal. But if you interpret
this in a fairly broad way, which is the case for many environmental
conventions, one can make an argument that is is included.

The official interpretation of the United States that Ambassador Richardson referred to does not exist as far as I know. What he said here today may be the most authoritative policy statement that has been made in this area in terms of his position and the specific topic that he was addressing. The only thing that I am aware of are statements before Congressional committees that if there is a significant risk posed to the environment, then the United States might interpret the London Convention as including and therefore banning dumping such sub-seabed disposal, unless there was further action taken by the IAEA and the parties. If, on the other hand, there is a low probability of releases over periods of concern, especially if the consequences are not severe, then apparently the United States position is that it is not legally affected and that it would require a completely different set of arrangements.

It is important to establish the degree of authority vested in the International Atomic Energy Agency for its oversight of radioactive waste disposal at sea. The IAEA gives advice to countries for their guidelines covering non-high level wastes. Countries are required to consider this advice as they enact their legislation and carry out their policies. For high-level wastes that are unsuitable for dumping at sea, which are to be defined by the IAEA, their advice to countries is legally binding. The 4,000 meter limit for dumping low-level wastes was accepted as a guideline at the IAEA. It was then sent back to the representatives of each country which is a party to the convention. In their meeting of October, 1979 this guideline was accepted as a resolution that there should be no disposal of non-high-level radioactive waste in water depths of less than 4,000 meters. This procedure brings even the guidelines for low-level wastes much closer to an actual legal requirement that is binding on every country that has ratified the treaty.

I would like to mention five major policy questions related to subseabed disposal. First is the international cooperative program on the research and development of seabed disposal. This is important for two reasons. First is that the United States has strong desire and responsibility to monitor what the other countries are doing with radioactive waste disposal at sea. Second, we have the responsibility to exchange data, information and scientists to ensure that the best possible scientific methods are used. There are no other real programs conducted at the international level in the radioactive waste management area, except for some cooperation between the United States and Sweden on disposal in granite. If the United States makes a finding, it is put to the scrutiny of five, six, or possibly now seven different countries. That I think is important.

The second policy question is the specific attitudes of countries that have strong influence over ocean pollution and nuclear waste disposal at sea. The Canadians and the Swedes are particularly interested in the annual international meetings in order to monitor what is happening. They want to know what is happening and they may even end up sharing information or joint cruises or experiments in this area. Other countries such as France and West Germany are participating to contribute and learn all they can. A third category of countries, especially the United Kingdom and Japan, has a direct need to consider the sub-seabed as a possible disposal option for high-level wastes. Japan now sends its spent fuel (and will continue through the 1980's) to Europe to be stored and event-

ually reprocessed. Eventually, they will either have the spent fuel
or the resulting high-level wastes returned to them. By the later
1980's Japan will be generating spent fuel that will be stored
locally. It is not at all clear where it will be disposed of. Although not so tightly, the United Kingdom is similarly constrained.

So the problem is interesting in two ways. One is that we do not
want them rushing out and making a conclusion before we do that
this is an interesting option and they are going to go dispose of
the stuff in the sub-seabed. The other is that at least the British
are much more interested because they are so constrained in their
options. It is very likely that Japan will either have to dispose
of it in some other national territory (if that turns out to be the
best option) or have to rely on some international area. One such
possibility is the sub-seabed.

The third policy question is the attitudes of the International
Atomic Energy Agency, the Nuclear Energy Agency, and other related
multilateral organizations. Both of those agencies are interested
in the sub-seabed disposal program. The Nuclear Energy Agency of
the OECO sponsors the cooperative research and development program.
They make facilities available and help publish the reports, especially to encourage cooperation and exchange of information. Each of
these agencies has to deal with a whole spectrum of opinions, so
their responses are not tailored by just one national audience.

As for my own personal response to the earlier question about less
developed countries and what they think about industrialized
countries using the so-called "common heritage" for dumping of high-
level waste disposal, I think that this is an important question and
I do not think that there is a real answer yet. It is very hard to
know. We have done some interviewing among some Indian officials,
but it is so preliminary that I do not want to say anything about
results because I do not trust them myself. There are lots of other
countries beyond India; for example Argentina, Brazil, South Korea,
Taiwan, that have large nuclear programs.

The results of the International Nuclear Fuel Cycle Evaluation (1977-
1980) constitute the fourth policy question. These point to a lot
of pressure on the United States to grant the importance of nuclear
power programs to other countries even though we seem to be losing
control of our own program. There are a couple of messages there.
One is that regardless of events in the United States, we are being
told that we are being too rough on nuclear power and that it has a
more important role than we understand. Whether this is correct or
not I will leave to your judgment. Second, there is some movement
in international cooperation on spent fuel management, which I am
particularly interested in from the point of view of slowing the
spread of nuclear weapons. Whether that will ever get into specific
activity such as storing spent fuel jointly on an island, such as
Palmyra Island in the Pacific, I'll leave for other speakers or for
you to think about.

If something like that does happen, and there are some non-prolifer-
ation reasons for trying to store spent fuel, especially from
countries such as Taiwan and South Korea, part of the agreement has
to be arrangements for the eventual disposal of active waste.

The last policy question I raise is the comparative assessment of
political acceptability. It is frequently held that land-based
disposal options are easier to implement somehow than sub-seabed
disposal, because sub-seabed disposal faces international political
problems. It is not yet clear that this is true because there are
several local, state and federal jurisdictional questions that make
land-based disposal options highly controversial. I do not know
what the final answers are, but it is clear to me that the seabed
disposal option avoids a very complicated U.S. legal picture except
for transportation and port facilities. It avoids the very complicated final siting issue within one of the states on the United
States. On the other hand, of course, it raises some new questions
of foreign policy and international law.

Actions by the United States in this area are important not only as
assertions of our policy but also as very strong influences on other
countries. It is not as much what happens in Sweden, France, or
Germany, for instance, that influences what happens in the United
States, as it is the other way around. Some of our decisions have
already been very influential in other countries.

OCEAN DISPOSAL OF RADIOACTIVE WASTES:
AN EXAMINATION OF U.S. POLICY TRENDS

Clifton Curtis
Attorney, Center For Law and Social Policy
Washington, D.C.

Today's discussion is timely. Given the increased attention that is
being devoted to resolving the "back-end" problems associated with
the use of nuclear energy, it is important that current knowledge
and future plans with respect to the "ocean alternative" be publicly
aired and examined. At present there exists a consensus within the
United States that the ocean alternative is not a viable near-term
disposal option for high-level radioactive wastes, assuming it were
legal. Similarly, ocean disposal of low-level radioactive wastes is
not considered by the U.S. to be an environmentally or economically
preferable disposal option at present. Absent highly acceptable
land-based respositories, however, the oceans will receive increasing
consideration as a disposal option -- if for no other reason than
their "out of sight - out of mind", limited constituency, appeal.
While scientists disagree on the degree of harm that has been or
could be caused by ocean disposal of radioactive wastes, there is
agreement that next to nothing is known about what the impacts from
low-level radioactive waste dumping heretofore have been on the
marine environment or on the food chain pathways.

My interest in this issue stems primarily from two broad concerns;
enhancing environmental protection of the oceans and increasing the
opportunities for public participation in our governmental decision-
making processes. As an attorney with the Center for Law and Social
Policy's International Project, I have represented environmental
organizations that are actively interested in the development of
sound ocean policies that provide for effective control of all
sources of pollution of the marine environment. That representation

has also been focused on opening up for larger and hopefully effective public participation the process of governmental decision-making;
seeking to ensure that the views of citizens are put before decision-
makers before government policy is determined.

Present U.S. policies and policy trends with respect to the ocean
disposal of radioactive wastes by this country fit roughly into
three categories. First, it is the current policy of this country
as a matter of practice not to use the oceans for the disposal of
any radioactive wastes. Second, regulatory standards are being
developed that would allow for ocean disposal of low-level wastes
pursuant to existing domestic law. In this connection, the United
States has advanced various positions in international gatherings,
primarily in connection with the London Dumping Convention, which
articulate standards that the United States believes should be
followed by those countries that are using the oceans as a low-level
radioactive waste disposal medium. Third, the United States is
presently engaged in extensive research to assess the technical and
environmental feasibility of using the ocean's seabeds as a disposal
medium for high-level wastes.

I would submit that present United States "policies" and federal
agency activities concerning the use of the oceans for disposing of
radioactive wastes can be characterized fairly as fragmented and/or
in an infant state. For example,

> the division of responsibility between EPA and the NRC
> for setting standards is uncertain;
>
> hypothesized release and transport events in the oceans
> have not been substantiated by actual studies and
> monitoring of past dumping activities;
>
> non-regulatory program activities are not well coordinated;
>
> technical positions presented at certain international
> meetings do not appear to square with other U.S. positions
> or with U.S. policies concerning matters impacting the
> global commons;
>
> the technical and environmental feasibility of seabed
> emplacement will not be known until the middle 1980's
> and program accountability pending go/no go decisions
> appears very limited; and
>
> NOAA's role and responsibilities for coordination and
> for research and monitoring are unclear.

Historically, a generally recognized U.S. policy that radioactive
wastes should not be allowed to pollute the marine environment dates
back to the 1950's. The 1958 High Seas Convention, to which the
United States is a party, provides that (Art. 25): "Every State
shall take measures to prevent pollution of the seas from dumping of
radioactive waste...." In October 1970 the Council on Environmental
Quality issued a report entitled Ocean Dumping: A National Policy
which concluded that ocean dumping presented a very serious and
growing threat to the marine environment. That report recommended
the policy of prohibiting ocean disposal of high-level wastes be

continued and that low-level wastes not be dumped except in very
limited circumstances where "no practical alternative offers less
risk to man and his environment."

In 1972, the Marine Protection, Research and Sanctuaries Act ("Ocean
Dumping Act") was enacted, prohibiting the ocean disposal, either
above or below the seabed, of high-level radioactive wastes and
regulating the manner and quantity of low-level waste disposal in
the oceans. In the fall of 1972, the United States was one of the
key instigating forces behind the adoption of the International
Convention on the Prevention of Marine Pollution by Dumping of
Waste and Other Matter ("London Dumping Convention"). That Convention, which has since been ratified or acceded to by the United
States and 42 other countries, prohibits the dumping of high-level
radioactives wastes (Article IV and Annex I), and allows the dumping
of all other radioactive waste only under special permits (Annex II).
Through that Convention, the United States has exercised considerable leadership in advancing environmentally responsible views on
various research, monitoring and other technical considerations
associated with the definition of high-level wastes and ocean disposal of low-level wastes.

In 1977, the current Administration introduced its National Energy
Plan, a plan which signaled its intention to develop a national
nuclear waste management policy and program. In the intervening
three years those proposed policies and programs have been the subject of extensive debate and review. The most recently completed
stage of review is contained in the March 1979 Report to the
President by the Interagency Review Group on Nuclear Waste Management. Many of the IRG findings and recommendations address broad-
ranging technical and institutional considerations and are therefore
applicable to ocean disposal considerations. Particular portions of
the Report discuss various considerations that will -- if adopted --
significantly influence the use of oceans as a disposal medium for
radioactive wastes. While it is expected that the IRG findings and
recommendations will almost all be adopted by President Carter, we
are still awaiting his pronouncement on nuclear waste management.
That statement -- which is long overdue -- will then serve as the
United States policy as this country seeks to resolve, in one fashion
or another, the problems associated with nuclear waste management.

In the interim, federal agencies have been involved in a variety of
programs which represent "policy" of a sort. For many of those
agencies, the IRG Report already represents the existing policy
statement that is influencing current and proposed activities.
Additionally, the positions that have been taken by U.S. representatives at international meetings have represented statements of U.S.
policy.

Other individuals, more technically specialized than I, have and will
continue to address the scientific considerations that are applicable
to the ocean option. In the remainder of my remarks, I have highlighted very briefly some of the considerations which should be
addressed in the context of U.S. policy or policy trends.

GENERIC CONSIDERATIONS

While the dominant focus of the IRG Report is on land-based geologic

disposal, consideration is given to the ocean disposal option, institutional responsibilities in relation to that option, general
proposed time frames for resolving the waste management problem, and
opportunities for public participation. As recommended by the IRG,
the primary objective of waste management planning and implementation
is that nuclear waste from all sources should be isolated from the
biosphere and pose no significant threat to public health and safety.
Since resolution of the waste management problem will not be made
overnight, it is expected that the Administration will develop an
interim strategic planning strategy, develop both programmatic EISs
and site-specific EISs prior to selecting particular technical
strategies, and pursue the development of criteria and standards
by DOE, EPA and NRC, in an accelerated manner.

The lack of coordination concerning federal activities and the unclear division of responsibilities amongst the agencies make it very
difficult to know what U.S. policies presently exist with respect to
the ocean disposal option. As noted in the IRG Report (at p. 87),
the resolution of institutional issues may well be more difficult
than finding solutions to remaining technical problems. Among
federal agencies, for example, there is a strong need for better
coordination of work effort on such matters as radioactive waste
research activities and standard setting. Concerning federal/state
and federal/regional relationships, the IRG preference for "consultation and concurrence" will need to be worked out in a manner that
ensures adequate state involvement in any site selection decision-
making process. This federal/state approach, including the creation
of the proposed Executive Planning Council, needs to be much more
precisely defined as to the participants' roles, rights, and responsibilities.

Another institutional aspect of nuclear waste management decision-
making process involves public participation. As a matter of policy,
the IRG recommended that additional and more effective means for
increasing public participation be required. With respect to the
ocean disposal option, this will require a much more publicly
reviewed and reviewable program than has been the case to date. For
example, the IRG recommendations would imply that research efforts
that have been or are being undertaken, by such agencies as the
Department of Energy, EPA, NRC, NOAA and the Navy's Office of Naval
Research, will need to be regularly publicized and subjected to
review that brings the various program pieces together.

At a broader level, the IRG recommended that DOE be given the primary
responsibility "for developing and integrating the overall planning
for the non-regulatory nuclear waste management program and for inter-
facing with the regulatory programs" (at p. 117), and for "developing and coordinating a national plan for LLW" (at p. 106). DOE's
lead agency responsibility will need to be handled in a manner that
objectively assesses the seabed option for radioactive wastes. As
part of that responsibility -- and especially critical to an evolving
policy -- would be DOE's biannual update of the waste management
plan simultaneous with updates on the National Energy Plan. As a
matter of IRG proposed policy, the IRG participants believe "that
the DOE can and must conduct the waste management program in a
responsible, careful and open manner and that, over time, DOE can
gain public confidence in its ability to do so" (at p.119). At
present there does not appear to be sufficient accountability by

DOE in performing such a lead agency function. Detailed progress
reports should be made available to the public on a regular basis,
with meaningful opportunities for reviewing and commenting on work
program goals, objectives and activities.

In this regard, it might well be appropriate for DOE to delegate to
the National Oceanic and Atmospheric Administration the lead agency
responsibility for coordinating the various non-regulatory ocean
disposal program activities. Pursuant to Public Law 95-273, NOAA
has responsibilities for coordinating and monitoring federal agency
activity as it relates to ocean pollution research and development
and monitoring, particular emphasis on the inputs, fates, and effects
of pollutants in the marine environment. In this context, NOAA's
first five-year plan, which in part established priorities for
national needs and problems, ranked the disposal of radioactive
waste as a high priority concern. Additionally, under Title II of
the Ocean Dumping Act, NOAA has specific responsibilities for monitoring and research regarding the effects of ocean dumping of radioactive wastes. To date, however, NOAA's involvement has been
limited and inadequate. It remains to be seen whether NOAA's future
activities will reflect the fact that research and monitoring with
respect to ocean disposal of radioactive wastes is a high priority
national need.

HIGH-LEVEL WASTE ACTIVITIES

In assessing the technical strategies for management and ultimate
disposal of radioactive wastes, the IRG focused most of its attention
on high-level and transuranic wastes. As to the interim strategic
planning basis for HLW, the IRG recommended that (at p. 61):

> "The approach to permanent disposal of
> nuclear waste should proceed on a step-
> wise basis in a technically conservative
> manner.
>
> "After having examined the status of knowledge
> relevant to disposal in mined repositories
> and by such other technical options as
> placement in deep ocean sediments, place-
> ment in very deep drill holes, placement in
> a mined cavity in a manner that leads to
> rock melting, partitioning of reprocessing
> waste and transmutation of transuranic
> elements, and ejection into space, we con-
> clude that near-term program activities
> should be predicated on the tentative assump-
> tion made for interim planning purposes that
> the first disposal facilities will be mined
> repositories. The nearer term alternative
> approaches (i.e., deep ocean sediments and
> very deep holes) should be given funding
> support so that they may be adequately evalu-
> ated as potential competitors. Funding of
> other concepts should allow some feasibility
> and preliminary design work to proceed. Once
> the NEPA process has been completed, program
> activities can be tailored accordingly."

This interim strategy is consistent with Strategy III of the IRG
Staff Subgroup on Alternative Strategies for disposal of high-level
wastes, which provides that (at p. 50):

> "for the first facility only mined reposi-
> tories would be considered. However, three
> to five geological environments possessing
> a wide variety of emplacement media would
> be examined before a selection was made.
> Other technological options would be con-
> tenders as soon as they had been shown to
> be technologically sound and economically
> feasible."

To date the Department of Energy's primary involvement with the
ocean disposal alternative has been through its Seabed Disposal
Program (SDP). The present primary goal of the SDP is to assess the
technical and environmental feasibility of disposing of high-level
nuclear wastes in the seabed of the oceans, with a secondary objective of developing and maintaining a capability to assess the seabed
nuclear waste disposal program of other nations. Currently, the SDP
is engaged in Phase II of a four-phase program, broken down as
follows:

> Phase 1 -- Estimation of technical and environ-
> mental feasibility on the basis of historical
> data. Completion date: 1976

> Phase 2 -- Determination of technical and
> environmental feasibility from newly acquired
> oceanographic and effects data. Estimated
> completion date: 1985-87

> Phase 3 -- Determination of engineering
> feasibility and legal and political accept-
> ability. Estimated completion date: 1993-95

> Phase 4 -- Demonstration of disposal facilities.
> Estimated completion date: 2000

Given that the SDP is several years away from completion of Phase 2,
it is far too early to assess the feasibility of this option. In
the event it is determined to be technically and environmentally
feasible, a programmatic EIS is planned for the early 1990s. Similarly, the program envisions waiting until after Phase 2 has been
completed to decide whether U.S. or international law should be
changed to accommodate this high-level waste disposal option. And
in this regard, it is important to keep in mind the fact that the
United States has established by law an unequivocal policy that the
seabed emplacement of high-level wastes is prohibited. Furthermore,
that same prohibition exists under the London Dumping Convention
absent guarantees that such seabed emplaced wastes would be completely isolated from the marine environment.

Leaving aside the issue of its legality, however, the recommendations
of the IRG provide a clearer focus for the relationship of ocean
disposal -- as a possible second generation nearer-term disposal
medium, -- to the preferred land-based disposal mediums.

Consistent with the IRG recommendation that the nearer-term alternatives be given the necessary funding support, the SDP is currently
receiving more than $5 million for FY 1980, with projected funding
increases that would provide at least $15 million by FY 1984. Given
the heavy infusion of research dollars, caution must be exercised to
ensure that the SDP remains neutral -- that it not serve as an advocate for use of the seabeds absent favorable Phase 2 and Phase 3
feasibility analyses, and absent the environmental impact analysis
that is currently expected in the early 1990's. Requiring DOE to
issue publicly reviewable progress reports for all non-regulatory
aspects of the waste management program as I mentioned earlier,
would contribute to a neutral SDP psoture. Obviously, such reports
and their public review and critique would be much more frequent
than the public participation opportunities that would occur at the
end of Phase 2 or Phase 3.

LOW-LEVEL WASTE ACTIVITIES

Since 1970, the U.S. policy has been to dispose of low-level radioactive wastes through shallow land burial. As indicated in the IRG
Report (at p. 77), a coordinated national program for management of
low-level wastes does not exist. As they found, "there presently
exists neither a coordinated national program for management of
these wastes nor an institutional mechanism to deal effectively with
these issues" (at p. 106). Given this problem, the IRG recommended
to President Carter that the Department of Energy assume

> "responsibility for developing and coord-
> inating the needed national plan for LLW
> with active participation and advice from
> other concerned Federal agencies and input
> from the States, general public, and industry
> in its formation."

While the IRG did not undertake a detailed evaluation of low-level
waste manangement, it did find that (at p. 78):

> "Research and development for improved methods
> of shallow land burial of LLW and of altern-
> ative methods of disposal should be accelerated
> because shallow land burial, as currently
> practiced, may not be adequate disposal method
> for all LLW in the future. Knowledge of the
> performance of shallow land burial and ocean
> disposal as presently practiced, is primarily
> empirical. Best estimates indicate that
> heterogeneity of the wastes, the extreme
> range of their physical and chemical proper-
> ties, and their interaction with the ground or
> ocean sediments after disposal are, at present,
> sufficiently complex as to make it difficult
> to confidently predict their long term behavior
> and their potential hazard to man. Improved
> and alternative disposal methods will be
> required to meet the growing needs of LLW
> management."

The Environmental Protection Agency is responsible for issuing a variety of environmental guidelines and standards applicable to ocean disposal of radioactive wastes. Pursuant to the Marine Protection, Research and Sanctuaries Act of 1972, EPA regulates the dumping of low-level radioactive waste in the ocean or beneath its seabed, in accordance with certain specified considerations. Consistent with the statutory mandates of the Act, EPA has promulgated regulations which prohibit the disposal of high-level radioactive wastes and require that all other radioactive materials must be packaged or containerized to prevent their dispersion or dilution in ocean waters.

By 1981, the agency expects to issue its siting criteria for ocean disposal of low-level wastes. By 1984, EPA is expected to establish specific sea disposal packaging and matrix container requirements as well as site-specific monitoring requirements for low-level wastes, with one ocean disposal site to be selected by 1985. Additionally, the IRG recommended (at p. 79) that by 1981, DOE and NRC should review existing and alternative low-level waste disposal techniques and determine whether any should be adopted in the near future. Given EPA's responsibilities in this area, that agency should be designated as a lead participant in making such decisions. Prior to undertaking these various activities and meeting the proposed schedules, however, there should be an initial decision made as to whether it is appropriate or feasible to give such serious consideration to the use of the oceans as a low-level waste disposal option.

It seems essential that a priority commitment be made to actual studies of past dumpsites before the U.S. gives serious consideration to the use of the ocean or its seabeds as disposal sites for low-level wastes. Of the various agencies, to date EPA appears to be the only agency which has engaged in any substantive research in this area. Its successful efforts using manned and unmanned submersibles on three occasions to survey deep sea radioactive waste disposal sites are commendable, but EPA and other agencies, especially NOAA, should significantly increase the number and scope of such studies in order to substantiate hypothesized radioactive waste release and transport events.

The Nuclear Regulatory Commission's involvement with the ocean disposal option for radioactive wastes has been limited. Under the Atomic Energy Act, as amended, and the Energy Reorganization Act of 1974, NRC has been given responsibility for ensuring that commercial radioactive waste management operations are performed in a safe and effective manner, setting standards that serve as the basis for licensing activities. The NRC has underway the development of a specific regulatory program for management of low-level radioactive wastes. It is developing proposed regulations for disposal of such wastes by shallow land burial by 1980 and is similarly intending to develop proposed regulations for disposal of such wastes by at least one alternative method by 1981. Among alternative disposal options recently reviewed pursuant to an NRC contract, however, ocean disposal concepts appear to be among the least viable disposal alternatives. (Evaluation of Alternative Methods for the Disposal of Low-Level Radioactive Wastes, prepared by Ford, Bacon and Davis, Utah, Inc., NUREG/CR-0680, July 1979).

Since the NRC's standards must follow EPA guidelines and implement
EPA standards, it will be helpful to better understand the regulatory relationship between the NRC and EPA. In this regard, the IRG
Report indicated that the relationship of the respective regulatory
schedules of the two agencies will be examined in a Memorandum of
Understanding, that is yet to be developed. That Memorandum should
be developed as soon as possible.

In the international arena, the United States, primarily through
EPA representation, has been an active participant in meetings which
have been convened pursuant to the London Dumping Convention. Since
the Convention became effective in August, 1975, four annual consultative meetings, as well as numerous intersessional technical
meetings, have been held to consider amendments to the Convention
and other matters. Following the United States' ratification of the
Convention, Title I of the Ocean Dumping Act was appropriately
amended to provide that

> "to the extent that he may do so without
> replacing the requirements of the (Act),
> the Administrator, in establishing or
> revising such criteria, shall apply the
> standards and criteria binding upon the
> United States under the Convention, in-
> cluding its Annexes." 33 U.S.C. S1412

U.S. positions and activities as they relate to actions that have
been or might be taken under the Convention have been subjected to
varying degrees of interagency and public review. Most of the
positions have involved recommendations that IAEA perform certain
work program tasks for the benefit of the Convention. In this
regard, the United States continues to encourage the accomplishment
of the following tasks, among others:

> (1) further review of the oceanographic model
> and assumptions that have been in use since 1973,
> given that the existing model (a) does not
> accurately represent the actual physical pro-
> cesses in the ocean, (b) it is unsuitable for
> evaluation of concentration levels for periods
> longer than a few hundred years, and (c) its
> failure to address many of the possible pathways
> for radioactive wastes through the environment;

> (2) detailed consideration of an administrative
> (compliance) mechanism as well as a technical
> performance evaluation program for waste isolation
> in order to support any release rate limit concept.
> Such a program should require, among other things,
> uniform packaging designs developed for different
> classes and types of wastes, and tested, both under
> simulated conditions and in situ, to establish
> acceptable standards of resistance to pressure
> corrosion, and leaching at depths in excess of
> 4,000 meters;

> (3) further consideration of radioactive releases
> (in addition to the impact of such releases on man)
> to include methods for assessment of localized impact

on sensitive elements of the marine environment
(including bioassay procedures for sublethal
effects) as well as the need to further reduce
those release rates to more adequately fulfill
the intent of Article I and II of the Convention
to prevent marine pollution;

(4)   development of improved risk estimation procedures including more specific recommendations
regarding environmental monitoring and food chain
pathway analyses;

(5)   establishment of a global limit on the number
of radioactive waste disposal sites, and an inventory of radioactive waste input to the sea so
that an estimate of the capacity of the marine
environment to accept radioactive waste from all
sources can be developed;

(6)   that the research programs and concepts
assessing the feasibility of seabed emplacement
of high-level radioactive wastes be applied to
low and intermediate-level wastes; and

(7)   that hard scientific data and measurements
begin to test and verify hypothetical models and
assumptions regarding the Northeast Atlantic
dumpsite, and that the required environmental
assessment include, for example, a thorough treat-
ment of the matters contained in Annex II of the
Convention and in Sections B.1.4 and B.1.5 of the
IAEA's August 1978 Recommendations concerning the
dumping of radioactive wastes.

It remains to be seen whether the United States views on ocean disposal of low-level radioactive wastes as expressed and presented at
past consultative meetings under the London Dumping Convention will
be adequately addressed.   As I have indicated, many of the U.S. concerns under the London Dumping Convention have requested the IAEA to
perform certain studies.   Absent IAEA action on these matters, will
the United States take upon itself the responsibility for addressing
them?   Alternatively, should not the U.S. press for the creation of
a standing international working group, similar in structure to the
Seabed Working Group that has been organized under the NEA's Radioactive Waste Management Committee?   Such a working group, set up
under the auspices of some international agencies responsible for
nuclear waste disposal matters, could address issues on which the
United States and other parties to the London Dumping Convention
have unsuccessfully sought IAEA review.

In 1977, the Organization for Economic Cooperation and Development
(OECD) Council adopted a Decision establishing a Multilateral Consultation and Surveillance Mechanism for Sea Dumping of Radioactive
Wastes.   That Decision, to which the United States is a party, is
especially significant since it represents the first time that a
mechanism has been established for formal review of ocean dumping of
radioactive wastes.   Pursuant to Article 2(a)(iii) of that decision,
a group of technical experts met in November 1979 to review the

continued suitability of the current low-level radioactive waste
dumping site in the Northeast Atlantic. Technical experts from the
United States participated in the meetings, concurring in the conclusions and recommendations that were adopted by the group. Among
other things, those experts concluded that:

> (v) There is a need to develop a site-specific
> model of the transfers of radionuclides, par-
> ticularly on short and medium time-scales, from
> the dump area to human populations. Therefore,
> there is clearly a need to continue investigations
> presently aimed at improving our knowledge of
> transport processes in the Northeast Atlantic. It
> is recommended that a well defined programme plan be
> developed over the next 12 months within the appro-
> priate international framework to meet this objective.

> (vii) As a result of their review of the suita-
> bility of the site, the Group of experts con-
> cluded that the site would be suitable for the
> receipt of packaged radioactive wastes during the
> next five years at annual rates comparable to those
> reached in the past. However, if current annual
> dumping rates were to be exceeded by a factor of
> ten, it would be desirable to reconsider the suita-
> bility of the site.

The report, once final review and editing has been completed by the
participants at the November 1979 meeting, will be sent to the OECD's
Environment Committee and the Committee on Radiation Protection and
Public Health for comment. The final report and comments will be
submitted to the NEA Steering Committee for Nuclear Energy meetings
in April 1980. At that meeting, U.S. participants will have the
opportunity to advise others present of U.S. "policy" with respect
to the decisions made by the Group of experts.

The well-defined program plan for a site-specific model of the transfers of radionuclides that is to be developed within the next twelve
months should be the focus of considerable U.S. attention and effort.
Many of the concern that have been raised at London Dumping Convention consultative meetings are directly applicable to the modeling
which the NEA has proposed. Given the concerns the U.S. has raised
at those meetings, and given the findings of the IAEA advisory group
contained in the "Criteria for Selection, Management and Surveillance
of Dumping Sites" (Jamaica, December 1978), I seriously question
whether the United States should support a 5-year extension of use
of the Northeast Atlantic dumpsite -- especially where the extension
permits annual dumping rates to exceed by a factor of up to 10 the
annual rates which have previously been allowed. Dissenting from a
decision to allow continued use of the site would also be consistent
with U.S. NEPA requirements with respect to actions affecting the
global commons, since the site suitability review document that came
out of the November 1979 NEA meetings would fail as an environmental
impact statement. Assuming the United States does not object to the
proposed extension as recommended by the Group of experts, such a
position would seem to reflect a decreasing interest on the part of
the United States in requiring adequate research and monitoring of
radioactive waste disposal practices.

CONCLUSION

In conclusion, the oceans -- which cover nearly three-fourths of the world's surface -- occupy a critical role in maintaining a livable environment. Given the extremely hazardous nature of radioactive wastes, their disposal into our oceans is fraught with potentially irreparable consequences. National policies that determine the manner in which we seek to protect, preserve, and utilize our vital marine resources must be cautiously and rationally advanced and subjected to continuing and rigorous review.

NATIONAL AND INTERNATIONAL POLICY
CONSIDERATIONS: A PANEL DISCUSSION

Edward M. Mainland
Department of State
Washington, D.C.

Well, if I had to articulate what the United States' policy has been and is in regard to radioactive waste dumping at sea perhaps I could begin by noting that our approach in international forums has been to express reservations about ocean disposal of radioactive waste in the sense of highlighting major questions that need answers. We ourselves do not dump at the present time; however, we recognize that some countries will continue to use this form of disposal. Indeed we ourselves are conducting an accelerated evaluation of the low-level waste option, and the expanding amount of wastes being generated may increase both the amounts deposited and the number of countries involved. The United States is therefore pursuing within the Ocean Dumping Convention, the IAEA, and OECD/NEA the goal of ensuring that such wastes as may be deposited in the oceans are handled in the environmentally safest way. Specifically, U.S. policy is to press for effective surveillance and consultation procedures, careful selection and limitation of dumping sites, an acceleration of research on the impact of emissions on marine eco-systems, international acceptance of the need to utilize isolation and containment of nuclear wastes based on the application of im-proved packaging techniques, and the development and implementation of comprehensive monitoring programs before, during, and subsequent to any disposal operations so that we can be aware of harmful effects at an early state. Now, in the IAEA this has resulted in the inclusion of the philosophy of isolation and containment and a minimum disposal depth of 4000 meters in IAEA Recommendations, based on an improved oceanographic model which still needs continued improvement. The IAEA's work program over the next several years will address many of the issues which we have posed at the London Dumping Convention meetings.

As a point of clarification, the IAEA ocean dumping Recommendations are not legally binding on dumping countries, but have acquired a little more force than mere recommendations as a result of a reso-lution by the Parties to the Ocean Dumping Convention last October at their Fourth Consultative Meeting which agreed "to implement the IAEA Recommendations to the best of their ability".

I might say also that the Mediterranean countries within the Barcelona Convention are considering an IAEA recommendation that would prohibit all disposal of radioactive waste materials in the

Mediterranean, based on the consideration that there is not sufficient depth there.

Now, with regard to the OECD/NEA Northeast Atlantic dumpsite, I
might just note that quantities being dumped there by four countries,
as pointed out in the historical background paper by Mr. Dyer, have
reached 150,000 curies per year. Some half-million curies or perhaps more have been dumped since operations began in 1967 at the
site. As Clif Curtis pointed out, if over the next four years an
order of magnitude increase were to be allowed, as NEA experts have
recommended, that 1.5 million curies could go in yearly, as I understand it, which would be a considerable increase over the entire
count during all the years since 1967; more radioactivity than has
been dumped everywhere since the birth of nuclear energy. Now, one
considerable problem that has come up is that over the years since
1967 there has been no monitoring of the site, no site-specific
analysis of effects. There hasn't been an adequate impact or risk
assessment done. The experts at a November 1979 NEA meeting finally
agreed on a site suitability assessment which they consider satisfactory. However, it is open to criticism for taking the approach
that lack of evidence means lack of risk, harm, or significant adverse effects. It can be argued that lack of information warrants
a more restrictive approach toward dumping than the experts recommended; until site-specific monitoring can verify models and hypothetical assumptions, it might be prudent to go a little slower
at the site. Also the situation raises an interesting question of
whether or not the United States might usefully advocate that the
NEA site suitability assessment embody something like the rigorous
standards of a United States environmental impact statement (EIS).
Certainly if the United States were to issue a permit for U.S. waste
generators and handlers to use the site an EIS would have to be em-
ployed. The U.S. cannot impose its own standards on others, of
course, but U.S. policy generally has been to encourage the world
community in the direction of environmental norms we believe justified.

With regard to high-level wastes I might just record a view that the
State Department has expressed to Congress several times, and which
Ambassador Richardson has reiterated, that sub-seabed emplacement
would be covered by the Ocean Dumping Convention and would be considered dumping and prohibited unless such emplacement could be
demonstrated not to pose a threat to the marine environment. Now
whether this threat is posed at any point can only be evaluated when
we have a more precise idea of the technical solutions for handling,
transport and emplacement of wastes, which we don't have now. But
were there a threat that in any emplacement process there could be
a release or discharge of high-level wastes into the water column,
the legal view would be that the Convention would apply. Regardless
of these legalisms -- and I recall Dr. Deese's point that probably
legal points won't be as important as the spirit of the Convention --
the mandate of the Convention is to prevent marine pollution. Legal
points should be measured against this overall purpose. More pertinent are political factors outside the realm of science or law. I
think that there is some evidence of the "don't put it in by backyard" attitude already at work among non-dumping countries. For
example, the OECD/NEA dump site, the original site, had to be moved
because of Portuguese objections that it was too close to their
territory. Similarly certain Pacific islands, I understand, have

already lodged objections with the U.N. to Japanese plans to conduct low-level radioactive waste dumping in the Pacific. Similarly, in 1976, there was a draft resolution introduced into the Hawaiian Senate which would have gone on record objecting to any sub-seabed emplacement in the area of the Hawaiian Islands. Another factor is that less developed countries are demanding access to the information that leads to decision-making that might affect them. For example, the Mexican delegation, at the last Ocean Dumping Convention meeting, made the point that Mexico has no access to OECD/NEA assessment documents or the documents that would allow an outsider to more adequately assess what is going on and participate in decisions that would possibly have an impact on the global commons. I might mention also that as a sub-seabed emplacement option is more thoroughly considered, certain countries' positions will have to be politically dealt with. For instance, the Mexicans oppose any dumping or disposal of radioactive wastes in the sea; the Soviets, in principle, oppose it; Scandanavian countries are on record as having opposed it; so there would appear to be considerable work to be done over the years in reaching an international consensus on a broader basis than just the small group of potential disposing countries which are now exchanging information in the OECD/NEA sub-seabed working group. An perhaps there is an analogy to local and state communities domestically in the United States which are to be part of the "consultation and concurrence" decision-making process governing United States domestic disposal programs.

Will there not be a similar need for wider consultation and acceptance by the international community, under the Convention or other instruments, in regard to disposal of long-lived, hazardous substances in the global commons where many nations may believe themselves to be potentially affected?

Dr. John Kelly
Center for The Study of Complex Systems
University of New Hampshire
Durham, NH

First, I would like to offer a general observation of the sub-seabed program, and then I would like to ask our speakers a few specific questions about their papers.

The sub-seabed program appears to be in a precarious position created by two prevailing attitudes toward waste disposal in this country; the "not-in-my-backyard" attitude and the "out-of-sight/out-of-mind" attitude. The "not-in-my-backyard" attitude is reflected in the large number of states which have passed legislation prohibiting the disposal of radioactive waste within their boundaries. As this opposition to land-based disposal increases, the "out-of-sight/out-of-mind" attitude may make the sub-seabed option appear more attractive. It is very important, however, that the political pressure created by these prevailing attitudes and the desire for a quick solution to the waste disposal problem not be allowed to distort the realities of the sub-seabed option. The sub-seabed program is still 15 years away from demonstrating its capacity to provide an acceptable means of waste disposal. On the other hand, it is equally important that land-based disposal not be selected just because the sub-seabed option is not immediately available, and that environmental concerns not be allows to foreclose the sub-seabed option before its feasibility has been fully tested. A rational approach to the issue is the only way to ensure that we arrive at a satisfactory solution to the problem.

Now I would like to ask Cliff Curtis a question. Cliff, you said that, "given the heavy infusion of research dollars, caution must be exercised to ensure that the sub-seabed program remains neutral." The sub-seabed program budget is about 5 million dollars out of a total budget of over 200 million dollars for all radioactive waste management. How do we exercise the same sort of caution for terrestrial disposal programs which are much more heavily funded? How do we maintain, or rather create, neutrality in land-based projects?

MR. CURTIS: I don't think that the exercise of caution with respect to government funded programs should be determined by the amount of dollars expended or authorized. Both the ocean disposal programs and land-based waste disposal programs should be cautiously developed and rigorously reviewed. As I mentioned in my earlier remarks, DOE's overall activities in addressing nuclear waste disposal options should be subjected to very close scrutiny and very regular progress reports whether it is looking at the ocean alternative or land alternatives. They need to take a neutral and objective position as they go through the economic, environmental and technical feasibility analyses.

DR. KELLY: That leads me to another question, if I may, which concerns your statement that the public should have a larger role in the decision-making process -- more "yes-no gates" as you put it. When you say "the public", do you mean the professional leaders of the environmental community who have the time to read the documents and raise the questions, or do you mean the general public? Would you also describe what you consider to be satisfactory mechanisms for promoting public participation.

MR. CURTIS: I would say it is both, that at different levels there are persons such as myself who benefit from the luxury of having access to some government officials who are intimately involved in the program. There are others who do not have that access but still have the general interest. When I was talking about public participation I also mentioned the phrase "public education" and I think that is an affirmative responsibility of the government in undertaking activities. It can be accomplished through progress reports, public meetings, briefings, and other efforts to lay before the populus their current and proposed activies. Through that education the public will become more informed and will play a more constructive and helpful role.

DR. KELLY: I would add scientists, like Dr. Hollister, to the list of those who share the responsibility for informing the public. I also think that, in addition to more "yes-no gates", the public should have an active role in the sub-seabed program. There should also be opportunities for interaction between the public and program scientists as a means of educating the public about the program and educating the scientists about the public's concerns.

The last question is for Dr. Deese. On page 10 of the notes that he circulated, he says that "given this combination of technical and

political constraints on possible final disposal sites" (and by
technical constraints he means the technical bias of research and by
political constraints he means overlapping jurisdictions) "governments are generally reluctant to specify their criteria for media
and site selection in any degree of detail. They fear that a systematic and precise set of criteria may raise difficulties with
options that they want very much to keep open. But they also need
more basic data from the laboratories and test sites before they can
be expected to perfect the narrowing process and the site selection
criteria."

It seems that you are saying that the government is caught in a hell
of a bind, a catch-22, in which they can't define site selection
criteria for fear of limiting their options, yet they cannot fully
explore their options until they define site selection criteria. The
government obviously can't have it both ways. Would you please ex-
plain how we might get out of this bind.

DR. DEESE: It is always much easier to identify the problem. The
British have taken a big step, I think, towards specifying criteria
before they actually started testing in an area. They have found
that local communities are not at all willing to start test drilling
without some idea of what would constitute acceptable final results.
The problem is that there is no guarantee for government decision
makers that such procedures will help, because we as social scientists
do not know if more information and public education leads to more
public acceptance or rejection. You can find examples of both.

DR. KELLY: That is all that I have for now.

         Alan Sielen
         U.S. Environmental Protection Agency
         Washington, D.C.

Much has been said throughout the day on the role of the United
States in various international negotiations and expert meetings.
Therefore, I think it would be useful to briefly take a look at the
vantage point from which the U.S. enters these international deliberations. While not necessarily disagreeing with what has been said
earlier, it might be useful to look at these issues from a little
different perspective.

As noted before, the U.S. looks at the question of ocean dumping low-
level radioactive wastes as a nation which does not dump these wastes,
but which is examining dumping as a possible option for the future.
Several nations whose policies we are trying to influence at the
multilateral and bilateral level have already concluded that ocean
dumping is an acceptable nuclear waste disposal option. Indeed,
some countries freely state that it is an essential part of their
overall national nuclear waste disposal strategy. Other countries
have concluded, often a priori, that the disposal of any nuclear
wastes in the oceans should be prohibited. Since the U.S. has not
yet reached any definite conclusions on the acceptability of sea
disposal, we are somewhat constrained in what we can advise others.

In general, a certain amount of healthy skepticism has been characteristic of the U.S. approach to this question. Our existing ocean

dumping regulations contain some guidance on what minimum criteria
would have to be met for the sea disposal of low-level radioactive
wastes to be considered. For example, regulations require that these
wastes radiodecay to environmentally innocuous materials within the
life expectancy of their containers and/or their inert matrix. This
philosophy of "isolation and containment" from the natural environment is a fundamental aspect of our approach to the question, and
one, incidentally, which has now been accepted by the IAEA in its
Recommendations on the dumping of low-level radioactive wastes. Unfortunately, I have not seen any evidence to suggest that the nuclear
waste dumping practices of other nations meet this very basic standard.

Recognizing constraints imposed by the present stage of U.S. thinking
on the subject, the U.S. can still encourage others to fully comply
with international requirements, and to bring a high standard of
scientific enterprise to bear on policy decisions. We can encourage
rigorous scientific and technical analysis of such questions as site
designation, monitoring the effects of dumping, and the actual dumping operation itself. Moreover, the U.S. has a large interest in
preserving the integrity of the environmental review process for
activies possibly having an adverse impact on any global commons
area.

The second area I would like to address relates to remarks made
earlier by Ambassador Richardson on the relevance of a Law of the
Sea treaty to ocean dumping issues. One theme that has pervaded out
discussion today is the dearth of meaningful information on environmental impacts associated with radioactive waste dumping. We have
heard of the IAEA oceanographic model not yet validated by hard
scientific data. We have heard that dumpers should comply with
international rules and guidelines established by the IAEA, the
OECD's Nuclear Energy Agency, and Annex III of the London Dumping
Convention. I think a basic problem is that there has been no effective legal means to ensure compliance with such international measures. There has been a serious void, in this regard, in existing
conventional international law. Clif Curtis made reference in his
paper to the 1958 Geneva Convention on the High Seas. If you look at
that document, and the other 1958 Law of the Sea Conventions, you
will find that there is only fleeting reference to the kinds of
marine environmental questions we must confront in the years ahead.
In the past decade there have been a series of global environmental
declarations and proclamations exhorting nations to behave sensibly.
But these are of a non-binding nature and provide little incentive
for strict compliance.

The comprehensive treaty emerging from the Law of the Sea Conference
would fill the present legal void and, among other things, obligate
states, for the first time, to conduct the environmental and scientific studies and programs we have been discussing today. In
addition to creating a sound jurisdictional framework for setting
and enforcing environmental standards, as noted earlier by Ambassador
Richardson, a Law of the Sea treaty would require states to do those
things necessary to make intelligent judgments on all aspects of
ocean use and management. States would have to develop contingency
plans for responding to accidental spills, monitor the effects of
pollution, and assess activities for potential impact on the marine
environment. Other measures emerging from the new law of the sea
environmental regime also point to a greater degree of accountability;

requirements to comply with international environmental rules;
compulsory procedures for settling disputes related to pollution;
and obligations regarding liability and compensation for harm to the
oceans.

A Law of the Sea treaty does have direct relevance to issues we are
examining today. Such an agreement could be the catalyst to trigger
the work needed to have a better understanding of the environmental
and scientific dimensions of the question of the sea disposal of
nuclear wastes.

ARTHUR TAMPLIN: The thing that concerns me about what I have heard
here today and, to a considerable extent, what I have heard over the
past 15 years associated with nuclear power and nuclear waste options
is that this area is perceived as a non-problem. Yet we are 35 years
into the nuclear era with no solution. There is a logical way to
conduct a program to develop a system by which one can adequately
dispose of radioactive wastes. However, not only this country but
every nation in the world has failed to proceed logically through
the necessary steps. The first thing that is needed to develop a
waste disposal system is firm criteria. These criteria have to be
substantial and capable of being validated either in the laboratory
or in field experiments. They have to be rigid criteria. You can't
build a barn until you know how big the barn must be. It is a
standing rule in any engineering program that before you begin the
construction you have to have the blueprints and you have to have
the specifications. There has been a total reluctance on the part of
anyone in the nuclear programs to firmly establish these criteria.
Once the criteria are established, then the research and development
program to develop the hardware and systems can begin. This R&D
program must have the purpose of demonstrating in the laboratory or
the field that the criteria are being met. The final step is a
demonstration program where it is all put together and demonstrated
that indeed the system does meet the predetermined criteria that it
was designed to meet. One is left in a "never-never land" unless
this is done.

We have in this country (and, alas, worldwide) no program for the
disposal of radioactive wastes, whether they are high-level wastes
or low-level wastes. If you look at what is happening in the United
States and elsewhere, you discover that they are proceeding backwards through this logical and necessary process for the development
of an adequate disposal system. First, a site is selected, then they
attempt to develop criteria which will allow for the site to be acceptable. In this country they choose the WIPP facility at Carlsbad,
New Mexico, their third attempt at a site in salt. This selection
was made in the absence of any criteria. At the present time, the
NRC has begun to develop criteria, and these NRC criteria would seem
to rule out salt as an acceptable media for the disposal of wastes.
But the DOE is proceeding along the same lines as its predecessor
agencies and moving backward through the logical process. It is
clear that the driving force behind the DOE efforts is the desire to
preserve the nuclear power option.

The nuclear power industry in this country and worldwide is in difficulty. In the United States they are being confronted with the
California Syndrome, where states are passing laws that you can't
construct any more nuclear power plants until you have demonstrated

an adequate waste disposal program.   Other states are adopting this.
A similar law was passed in Sweden.   The second problem confronting
the nuclear industry is that the real cost of nuclear power must include the cost of waste disposal.   The Public Service Commissions and
Public Utility Commissions are beginning to call for the utilities to
put the cost of waste disposal in the rate structure and do it now.
The third problem is that the utilities are becoming constipated.
Their spent fuel pools are beginning to fill up and they are going to
have to build some new pools on their sites or the government is going
to have to shut their reactors down.   These problems are the driving
force behind the DOE waste disposal program.   The driving force is
this desire to preserve the nuclear power option.

I would like to say just one final thing, and this is related to
something Ambassador Richardson discussed.   This is related to the
decision-making process for developing the criteria.   The process, as
I stated earlier, has really been backwards.   The criteria has been
adopted to accommodate the industry rather than the public health and
safety.   This clearly seems to be the approach that is occurring with
the disposal of radioactive wastes.   It was stated here that a 10-fold
increase in allowable ocean dumping is being proposed.   These proposals are made by international agencies composed of individuals with
a vested interest in the nuclear industry.   For example, the National
Commission on Radiological Protection and the International Commission on Radiological Protection include the same people.   They are
mostly people who have a vested interest in the radiation exposure
by industry.   In the ICRP they go to the international meetings, they
make recommendations, they go back to their own countries, take off
one hat and put on another hat and accept the recommendations that
they have just made. And in making their recommendations for exposure of the population at large, or for occupational exposure, they
have always looked over their shoulders to see what the nuclear power
industry could accommodate rather than setting rigid criteria that
would force the engineers who were designing these systems to come
up with systems that would meet the appropriate criteria when judged
against public health and safety.

> Richard Norling
> Staff Director
> Sub Committee on Oceanography
> Washington, D.C.

A couple of earlier speakers have mentioned the possibility that
seabed disposal just might get adopted regardless of scientific work
because it is so appealing to get this stuff "out of sight, out of
mind."   That is to some extent a matter for deep concern.   Back in
the 1950's and 1960's, before we were very conscious of what we were
doing, we were not only dumping drums of radioactive waste into the
ocean, we were loading up old Liberty ships with ammunition and other
munitions that were too dangerous to use, and sinking them.   The
government actually sank the defective reactor from a nuclear submarine.

The reaction to that profligate and unthinking use of the oceans for
waste disposal was passage by Congress in 1972 of the Marine Protection, Research, and Sanctuaries Act, or the Ocean Dumping Act, which
does clearly, in my opinion, make it totally illegal to dump high-
level radioactive wastes or to place them in the seabed.   The question

then arises, under what circumstances would that law be changed, and
would the pressure of state and local governments saying that "we
don't want it in our state or our locality" be enough to get that law
changed? Luckily, from my point of view, there are also members of
the House and Senate who view part of their charge as protecting the
oceans and the Ocean Dumping Act, which would have to be amended to
legalize the seabed emplacement of high-level radioactive wastes.
Jurisdiction over the Ocean Dumping Act is in the Oceanography Subcommittee and the Fisheries and Wildlife Conservation and the Environment Subcommittee of the Committee on Merchant Marine and Fisheries
in the House, and in the Senate it is the National Ocean Policy Study
Subcommittee of the Senate Commerce Committee. So I believe that we
need not fear an unduly speedy political decision that seabed emplacement would be the easy way to get out of this problem. Luckily,
we have an institutional structure where the hard questions will be
asked before the political votes will be taken on the issue.

Now under what conditions might I, as a subcommittee staff director,
recommend to the members of our Subcommittee that they repeal the
prohibition? It is my opinion that when man intervenes in the environment in any major way the amount of care that we exercise should
be commensurate both with the magnitude of the effects the interven-
tion would create, and with the length of time that those effects are
expected to last. In most cases that we consider as political issues,
we only have to consider the first criterion, how large an impact are
we going to create by doing x, y, or z. But when you are talking
about elements that are radioactive with half-lives of thousands of
years, it seems to me that the time span is more important a thing
to consider than the magnitude of the temporary effects.

When the time span is considered, it becomes difficult to choose
locations, let alone whether or not to accept the effects. We are
seeing in our subcommittee, and I am sure other people are seeing as
well, far more competition for space in the ocean to condust activities that previously were conducted on land. There is more and more
demand for fishing activities because meat prices are going up, and
the other sources of protein are becoming cramped because of the increasing world population. There was even talk of floating nuclear
power plants not too long ago. A lot of large companies are getting
ready to try ocean thermal energy conversion, biomass conversion,
current energy, wave energy, all of which would take space in the
ocean; and there is still a lot of shipping and it is still increasing, worldwide. There are many activities for which people are
starting to look to the ocean. When you pick a site that is going
to be taken for a very long period of time, one has to consider what
we are going to have to forego in the future by taking that site and
allocating it perpetually, in comparison to our life spans, for a
particular use.

In terms of making an actual decision on the scientific evidence, I
would recommend to the members of our Subcommittee that they require
an extremely high level of confidence, much higher than is usually
required by scientists in making predictions related to public policy,
primarily because of the long half-lives that we are talking about.
The second thing that I would require is monitoring, absolutely.
There have been some attempts in the United States, but not so much
by the other countries who previously have dumped things in the
ocean; we need to keep track of what is there and what it is doing.

The third thing would be the ability to recover the material and
relocate it.   We are talking about things that are going to be around
for so many human lifetimes, it seems to me that recoverability has to
be one of the criteria.   If we make a mistake, will somebody in the
future be able to correct it, or is it going to be perpetual?   I
believe we cannot afford to make a decision which is irreversible.

QUESTIONS AND ANSWERS SESSION

MR. ROOSEVELT: Thank you very much.   I would now like to encourage
the panel to continue discussion between themselves further for about
the next five or ten minutes and we will then turn to the audience
for questions.

DR. KELLY: I have a question for Dick Norling.   On the one hand, we
see that there are forces which may make the sub-seabed appear to be
a favorable option.   On the other hand, we might very well find ourselves in a situation where the terrestrial sites are unacceptable
and the sub-seabed site is also unacceptable because it can't meet
one of the criteria that you have stated.   For example, Dr. Hollister
told us that there are going to be certain areas where you just can't
find a definite answer and a certain amount of uncertainty the public
is going to have to accept.   Retrievability, for example, seems to be
technologically feasible but it could be troublesome.   What happens
if we find ourselves in the unsatisfactory situation of finding none
of the sites acceptable, neither the terrestrial nor the seabed sites?
Even if we shut down every nuclear power plant that is operating today, we would still have existing wastes which are improperly, temporarily stored.   How do we find our way out of this maze?

MR. NORLING: Well, I don't want to be a "doom-sayer", but we may not
be able to.   I believe the criteria that I outlined for consideration
of ocean disposal are also reasonable criteria to use in evaluating
other methods of disposal or storage.   I guess you lose some things
and you gain some things, depending on the way the government is
organized.   In Congress there is not a single committee charged with
considering and weighing the risk of all the disposal or storage
options.   I would almost argue that high-level radioactive waste is
not really disposed of unless you transmute it.   Because it is going
to be around for so long, we are really talking about storage.

There is one more reason why the law might not get amended even if
the scientific weight of evidence said that the most reasonable risk
were for ocean disposal, and that is that innovation moves upward
from the states--localities to states, states to nations, and then to
international law.   Not always, but that is the usual pattern.   In
this case, the treaty--to the extent that it prohibits high-level
radioactive wastes from being emplaced in the seabed--would have to
be repealed first, because the United States as long as is a signatory to the treaty has a legal obligation to enforce its terms.   I
assume, for instance, that the Justice Department would come in and
oppose any bills strictly on that legal position if we were required
by the treaty to keep enforcing a substantive provision against the
dumping.   So that could be an additional institutional and legal
barrier to the seabed emplacement option.

DR. KELLY: I have another question, if I may, for Cliff Curtis. You implied in your statement that there may be some problem in the fact that, according to the sub-seabed program, not until after Phase 2 are national laws and international laws to be amended to permit sub-seabed disposal of radioactive wastes. We have just heard from our colleague on the staff of the Oceanography Committee that there seem to be many legislative impediments to the sub-seabed option. Are scientists, like Dr. Hollister and crew, off on an errant mission, is it likely that they are going to conduct these millions of dollars of research only to find that there is absolutely no way in God's political world that we are going to have sub-seabed disposal of radioactive waste? Should amending relevant legislation be moved forward before we continue to support these scientists on their ships out in what Dr. Hollister calls "that boring part of the world"?

MR. CURTIS: It would seem to me to be an unwise and inefficient use of Congress' or the London Dumping Convention parties' time and resources to engage in a debate over the appropriateness of allowing the ocean seabeds to be used as repositories for high-level radio-active wastes until we have considerably more knowledge about the oceans and our technological capabilities. Once the Department of Energy has completed the Phase II level, there will be plenty of time to assess any legal matters. I appreciate the concern that you articulated on behalf of scientists whose work efforts might then be frustrated by the laws not being amended after their analyses have concluded that the option is valid. However, if seabed disposal of high-level radioactive wastes was found to be environmentally feas-ible then the men and women in Congress should act responsibly and consider it in comparison with other alternatives and arrive at some decision as to whether or not it is a viable option.

MR. ROOSEVELT: I think at this point we will go to questions or comments from the audience. I would repeat that you please indicate to whom you wish your question addressed, and I would like that person to repeat the question for the purposes of transcription. Are there any audience questions?

QUESTION: To Dr. Deese --

DR. DEESE: There are some interesting implications of that question. First, the London Dumping Convention had nothing to do with Vienna, but was formed with the cooperation of the Intergovernmental Maritime Cooperative Organization in London. With reference to specific legal interpretation of the Convention, I can only rely on my own writing, reading and discussion with many lawyers. It is not at all clear to me how you and Rich Norling reach the conclusion that sub-seabed disposal is clearly covered by, and banned as dumping by this treaty. I gave you my interpretation of the different ways you can look at the working. Beyond that I do not know what I can say. I have no interest or position at stake to protect.

MR. NORLING: I am not a legal expert on the London Dumping Conven-tion. I have not read it recently. But I do know what the Ocean Dumping Act requires. The point I made about the impediment to re-pealing United States law, or changing United States law, before a treaty is changed really does not prejudge whether the London Dumping Convention bans seabed emplacement or not. If the treaty does prohibit seabed emplacement, then the problem I was discussing does exist.

QUESTION: I guess I would like to direct this to several members of the panel but Mr. Curtis and Dr. Deese did specifically refer to somewhat of a degree of diffuseness in the interface between state and federal prerogatives in the definition of waste disposal. I guess my interpretation of that is that it comes out of the fact that the federal dollar is responsible for a great deal of research in the non-regulatory area but the legislative leaders have not given any details on the relationship between state and federal governments, their respective authorities and/or responsibilities. I would like that clarified, and any suggestions on how this interface could be better defined.

MR. CURTIS: In my earlier remarks, I mentioned that the inter-agency review group recommended a cooperation and concurrence approach for accommodating states' interests. In a sense the concurrence approach provides that states would have a veto right over land-based disposal options within their boundaries, but it was the hope and tenor of the IRG report that through this cooperation and concurrence effort they would not have to reach the loggerhead of being told by a state "No, you can't come in here." By providing for state participation in the development and review of research and technical feasibility analyses, hopefully they would be accepting the final decision. Regionally that same approach applies. Additionally, members of the proposed Executive Council that the IRG recommended involves representatives with different viewpoints, different political persuasions, different levels of our political system. They should be able to speak for these different groups and yet still come to a consensus. I am waiting to see how that comes together; how they actually put "meat" on those concepts.

MR. ROOSEVELT: Do any other panel members want to deal with that?

MR. TAMPLIN: My feeling is that if you are looking at either the land-base disposal or the seabed disposal of radioactive wastes and strictly on the state and nation issue, that the state should have no right to veto because the radioactive waste is not protected by state boundaries. If one state wants to accept a waste disposal facility that is totally inadequate, the radioactive wastes will migrate from there in the ground water into the neighboring states. Let's say that the state of Minnesota was only concerned about its own state, then it could allow a radioactive waste dump in the southeast corner of Minnesota. They wouldn't have to worry about it at all, anything that went into the river would go on down the Mississippi River and affect other states and leave them totally alone. So this has to be a decision that has to be made on a national basis. The difficulty that we are having right now is that we don't have a waste disposal program, we have no criteria. When you look at what the Department of Energy is doing, it seems that they want to dig a hole and throw the stuff in there. And this is what has all the governors and citizens upset. Once they get a good program and get some good criteria and get some confidence behind them, then I think this veto issue will fall out of the way.

DR. KELLY: The problem is how to align political considerations with scientific and technical criteria. I would prefer to have prospective sites selected on objective scientific and technical criteria rather than political acceptability. The fact is, however, that political expedience is leading the government to focus on federally-held land, such as the Nevada Test Site and the Hanford Reservation, rather than on sites selected on the basis of purely scientific and technical criteria. As a result, we are being asked to accept less than the optimal solution to the problem. What we need to do is involve the public in the site selection process so that they realize that it is in their interest to select the best possible site on scientific and technical criteria rather than political expedience. The "not-in-my-backyard" attitude must give way to reason.

MR. TAMPLIN: With respect to that, it is true that the Department of Energy is now investigating different media besides salt and they are doing it on government reservations where they presently have radio-active wastes stored. So far as the scientific acceptability of that goes, it is ridiculous. It is totally a political decision because in May of 1966 after 10 years of study the Committee of the National Academy of Sciences said none of those sites where waste is presently stored by the Atomic Energy Commission is in a suitable geological location for the disposal of any but the very dilute liquids. So nothing has happened since 1966 to change the nature of those sites, and this is after 10 years of study. Well, the AEC suppressed that 1966 report and it finally took Senator Church to force it down on the floor of the Senate in 1970. Just to show you how it works, the committee that prepared the report was disbanded and a new committee was appointed, with no overlapping membership. So these things are political.

DR. DEESE: I just wanted to comment on research money and veto power for the states. I agree strongly that radioactive waste management is a national issue; but there is too much power in favor of the federal government because the states generally do not have the people, the expertise, the access, the time, and the research dollars to influence the process. They are in a situation of being called on to make some important decisions, yet some either do not want to or cannot make such choices.

There is also a very complex set of Congressional committee jurisdictions. Some of them have real jurisdiction; many do not have any. The IRG did go a long way in resolving some of these problems. It was really an important step, if the results are implemented. But even if they are implemented, I do not think you can have the states do what some of us would like them to do. They just do not have the resources.

MR. ROOSEVELT: If there are no further questions, I would like to thank the two people who made the presentations and the remainder of the panel for your participation today. I appreciate it very much.

8. EPILOGUE

Michael J. Herz
Executive Vice President
The Oceanic Society

The year between the Oceanic Society's Public Policy Forum and of these Proceedings contained a number of significant events relevant to ocean disposal of nuclear waste. Because the conference and the publication resulting from it were designed to bring together much of the information from this field, we feel this update is an appropriate addition to the volume.

In his paper in this volume[1] Robert Dyer of the Environmental Protection Agency made reference to surveys conducted at the Atlantic and Pacific dumpsites between 1974 and 1978. Although the results of some of these operations had been published or made available as interim or final reports from EPA, the findings of the more recent surveys of both the Atlantic and Pacific had not been released. During the spring and early summer following the forum, local state, and federal officials, primarily from California, began to exert pressure on EPA for release of the documents. At the urging of Representative John Burton (Democrat, San Francisco) field hearings of the House Subcommittee on Environment, Energy, and Natural Resources (of the Committee on Governmental Operations) were scheduled by the chairman, Representative Toby Moffett (Democrat, Connecticut). Shortly after this announcement, in early September, the EPA released copies of 13 reports[2-14] on the Atlantic and Pacific surveys to California state officials.

The Congressional hearings, which occurred the next month,[15] focused on the results of the EPA research and on the roles of the EPA, the Department of Energy, and the Nuclear Regulatory Commission, the principal federal agencies concerned with nuclear waste and its environmental impacts. Analysis of the survey data performed by an ad hoc scientific committee created by the Oceanic Society to offer technical advice to the Congressional subcommittee was in essential agreement with EPA and California officials in concluding that there is no evidence to suggest that the Farallon (Pacific) site now poses a health hazard. However, all of these representatives also agreed that there was a need for future monitoring of this and other sites. EPA representatives pledged to develop an interagency agreement with the National Oceanic and Atmospheric Administration which would make possible a joint field monitoring operation. California officials also indicated that they planned to conduct their own surveys and that they would coordinate their activities with the federal agencies.

Because of poor record-keeping during the 1946-1970 period, when the U.S. was actively engaged in ocean dumping, the hearing also focused on attempts to document the contents and locations of nuclear dumpsites. Nuclear Regulatory Commission and EPA officials differed in their reports of the number of sites located off the Atlantic, Pacific, and Gulf coasts, but their totals equaled only slightly more that half

of the number documented in a report presented by the Committee to
Bridge the Gap, a small private-sector group which has uncovered
records which indicate that the total exceeds 50 sites. Also discussed was the recently published report[16] that the navy is considering disposing radioactive power plants of decommissioned nuclear sub-
marines at sea over deep ocean bottom areas which have been surveyed
in both the Atlantic and Pacific.

Discussions with the subcommittee staff at the close of 1980 indicated
that plans for the joint EPA/NOAA monitoring program were proceeding
well and that surveys relating to this and other NOAA programs might
cover all major dumpsites on the East, West, and Gulf Coasts. It was
also reported that EPA had requested all available information regarding dumping from the Department of Defense, which had, in turn, in-
structed all services to comply. Although not yet scheduled, the
subcommittee planned to hold East Coast hearings in 1981.

In November a hearing was held in Washington by the Subcommittee on
Oceanography of the House Committee on Merchant Marine and Fisheries,
chaired by Representative John Murphy (Democrat, New York). This
subcommittee examined the technical, environmental, and legal issues
surrounding the possible disposal of high-level wastes in the seabed
and also investigated current monitoring of the effects of previous
ocean dumping of low-level wastes. Representatives from the State
Department, EPA, NOAA, and DOE (and its contractors, Sandia Laboratories and Woods Hole Oceanographic Institution) presented information
on the status of the Subseabed Disposal Program for U.S. commercially
generated high-level waste, and regulatory controls, both national
and international, which will influence it. Discussion of the sub-
seabed planning revealed that a multifaceted research program designed
to determine its feasibility has been in operation since 1974. While
much of the effort is directed at the technical, environmental, and
engineering questions, legal and political issues are also being
weighed as well. While there appear to be differing opinions among
participating agencies and contractors, consideration is beginning
to be given to when and to what degree public participation will be
solicited.

There was also discussion of past U.S. nuclear waste disposal practices
and policies with attention being focused on the fact that documentation
on the location and contents of dumpsites is extremely poor or, in a
number of documented cases, absent. Representative Glenn Anderson
(Democrat, California) presented evidence that indicated that in addition to the 47,500 containers of low-level waste dumped near the Farallon Islands there was also a number of larger containers, at least
one of which contained a beryllium nuclear warhead. Other information,
some of it presented by Representative John Burton at the earlier hearing and some that was recently revealed by a retired navy pilot, suggests
that the navy and Air Force Reserve may have participated in previously
undocumented aerial dumps of low-level containers off both the Atlantic
and Pacific coasts.

Attention at this hearing also focused on the proposed ocean dumping
by the Japanese government of approximately 10,000 containers (500
curies) of low-level waste 500 miles north of the Mariana Islands.

There are also indications that if the Japanese achieve technological success and if world opinion is not adverse, they will enter into a more ambitious program of dumping several-orders-of-magnitude higher curie counts of low-level waste each year. However, public opinion, at least in the Pacific basin, appears to be strongly opposed to the plan. The governments of a number of Pacific entities (Papua, New Guinea; Guam, the Northern Marianas) have expressed great concern over the impact of the Japanese disposal, and some have gone so far as to formally request that the plan be dropped.

The events of 1980, beginning with the Oceanic Society forum, served to focus international public attention on past, present, and future aspects of the ocean disposal of nuclear waste. If the forum, and these Proceedings, have helped to make policy decision-makers, elected officials, and the public more aware of past and potential impacts of radioactive waste disposal practices, they have accomplished our objectives.

Bibliography

1. Dyer, R. S., "Sea Disposal of nuclear Waste: A Brief History,"
   this Proceedings, 7 pages.
2. Colombo, P., R. M. Neilson, Jr., and M. W. Kendig, "Analysis
   and Evaluation of a Radioactive Waste Package Retrieved from
   the Atlantic 2800-Meter Disposal Site." September 1978.
3. Dayal, R., I. W. Duedall, M. Fuhrmann, and M. G. Heaton,
   "Sediment and Water Column Properties of the Farallon Islands
   Radioactive Waste Dumpsites." September 1979.
4. Dexter, Stephen C., "Onboard Corrosion Analysis of a Recovered
   Nuclear Waste Container." Technical Note ORP/TAD-79-2, August
   1979.
5. Interstate Electronics Corporation, "Operations Report--A Summary
   of the Farallon Islands 500-Fathom Radioactive Waste Disposal
   Site." U. Environmental Protection Agency, Technical Note
   ORP-75-1, December 1975.
6. ------, "Operational Plan, Phase I, 1977 Farallon Islands Survey."
   IEC 446SP 550.
7. ------, "Operational Plan, Phase II, 1977 Farallon Islands Survey."
   IEC 446SP 551.
8. LFE Environmental Analysis Laboratories, "Radiochemical Analysis
   of Samples from the 900-Meter Pacific Dumpsite." September 1979.
9. Reish, Donald J., "Survey of the Benthic Invertebrates Collected
   from the United States Radioactive Waste Disposal Site Off the
   Farallon Islands, California." August 1978.
10. Robison, Bruce H., "Cruise Report: Farallon Islands Disposal Site
    Survey, Phase I, 25 August to 2 September 1977."
11. ------, "Midwater Trawling Summary: Farallon Islands Disposal Site
    Survey, 1977."
12. Rego, Jennifer A., "Deep-Sea Echinoids and Asteroids of the North-
    eastern Pacific: An Aid in Selecting Candidate Species for Chromo-
    somal Analysis-and-Observations Concerning Three Species of Sea
    Stars Collected by the Velero II." March 1980.
13. Schell, W. R., and S Sugai, "Radionuclides in Water, Sediment and
    Biological Samples Collected in August-October 1977 at the Radio-
    active Waste Disposal Site Near the Farallon Islands." July 1978.
14. Silver, Gary R., "A Taxonomic Review of the Farallon Island Sponge
    Fragments." February 1979.
15. Carter, L. J., "Navy Considers Scuttling Old Nuclear Subs." Science,
    1980, 209: 1495-1497.
16. Hearings on Ocean Dumping of Radiocactive Waste, before the Subcom-
    mittee on Environment, Energy, and Natural Resources of the Govern-
    mental Operations Committee. 96th Congress, Second Session, 1980
    (October 7, 1980, San Francisco, California).
17. Hearings on Radioactive Waste Disposal in the Oceans, before the
    Subcommittee on Oceanography of the Committee on Merchant Marine
    and Fisheries. 96th Congress, Second Session, 1980 (November 20,
    1980, Washington, D.C.).